WILKS · Mathematical Statistics

WILLIAMS · Diffusions, Markov Processes, and Martingales, Volume I: Foundations

ZACKS · The Theory of Statistical Inference

Applied Probability and Statistics

BAILEY · The Elements of Stochastic Processes with Applications to the Natural Sciences

BAILEY · Mathematics, Statistics and Systems for Health

BARNETT and LEWIS · Outliers in Statistical Data

BARTHOLOMEW · Stochastic Models for Social Processes, *Second Edition*

BARTHOLOMEW and FORBES · Statistical Techniques for Manpower Planning

BECK and ARNOLD · Parameter Estimation in Engineering and Science

BENNETT and FRANKLIN · Statistical Analysis in Chemistry and the Chemical Industry

BHAT · Elements of Applied Stochastic Processes

BLOOMFIELD · Fourier Analysis of Time Series: An Introduction

BOX · R. A. Fisher, The Life of a Scientist

BOX and DRAPER · Evolutionary Operation: A Statistical Method for Process Improvement

BOX, HUNTER, and HUNTER · Statistics for Experimenters: An Introduction to Design, Data Analysis, and Model Building

BROWN and HOLLANDER · Statistics: A Biomedical Introduction

BROWNLEE · Statistical Theory and Methodology in Science and Engineering, *Second Edition*

BURY · Statistical Models in Applied Science

CHAMBERS · Computational Methods for Data Analysis

CHATTERJEE and PRICE · Regression Analysis by Example

CHERNOFF and MOSES · Elementary Decision Theory

CHOW · Analysis and Control of Dynamic Economic Systems

CLELLAND, deCANI, BROWN, BURSK, and MURRAY · Basic Statistics with Business Applications, *Second Edition*

COCHRAN · Sampling Techniques, *Third Edition*

COCHRAN and COX · Experimental Designs, *Second Edition*

COX · Planning of Experiments

COX and MILLER · The Theory of Stochastic Processes, *Second Edition*

DANIEL · Biostatistics: A Foundation for Analysis in the Health Sciences, *Second Edition*

DANIEL · Application of Statistics to Industrial Experimentation

DANIEL and WOOD · Fitting Equations to Data

DAVID · Order Statistics

DEMING · Sample Design in Business Research

DODGE and ROMIG · Sampling Inspection Tables, *Second Edition*

DRAPER and SMITH · Applied Regression Analysis

DUNN · Basic Statistics: A Primer for the Biomedical Sciences, *Second Edition*

DUNN and CLARK · Applied Statistics: Analysis of Variance and Regression

ELANDT-JOHNSON · Probability Models and Statistical Methods in Genetics

continued on back

FOURIER ANALYSIS OF TIME SERIES: AN INTRODUCTION

FOURIER ANALYSIS OF TIME SERIES: AN INTRODUCTION

PETER BLOOMFIELD

Princeton University

RICHARD THORPE

John Wiley & Sons

New York • London • Sydney • Toronto

Library of Congress Cataloging in Publication Data:

Bloomfield, Peter, 1946–
    Fourier analysis of time series.

  (Wiley series in probability and mathematical statistics)
    1. Time-series analysis.   2.   Fourier analysis.
I.   Title.

QA280.B59 1976   519.2'32   75-34294
ISBN 0-471-08256-2

Printed in the United States of America

10 9 8 7 6

TO MY FAMILY

PREFACE

There are two groups of users of an applied book on time series analysis such as this. The first consists of students (undergraduate or postgraduate) who encounter a course on time series in their study of statistics or its allied fields. The second consists of workers in the many fields in which time series data arise. The fact that this book was written with both groups in mind has imposed noticeable constraints on the contents and presentation, but they have proved entirely beneficial.

In the interests of the second group, the statistical level of the presentation has been kept low. The minimum statistical knowledge needed to follow the essential sections corresponds to a single introductory course in statistics. Greater knowledge of statistics, combined with some experience in the analysis of observational data, would of course allow the reader both an easier passage and the opportunity for greater gain on the way.

The interests of students are best served, at least on their first contact with time series, by tying the presentation to examples. All the methods described in this book are introduced in the context of specific sets of data, so that the motivation behind a method is evident as it is developed. The abstract properties of a procedure are discussed only when the motivation has been solidly established.

Many people have difficulty when they first encounter Fourier analysis or the Fourier transform. The discrete Fourier transform is described in Chapter 3 and is used in one form or another through most of the remaining chapters. It is elementary from a mathematical point of view, involving nothing more advanced than the summation of finite series, not even calculus. However its properties are analogous to those of more difficult types of Fourier transforms. Careful study of Chapter 3 and the expenditure of some time on its exercises will convince the most faint-hearted that Fourier transforms can be fun! Complex numbers are used extensively in deriving the properties of the discrete Fourier transform.

However this is done purely for the notational simplicity, and to follow the algebra it is necessary only to know the rules for obtaining the sum and the product of two complex numbers.

The topics discussed are as follows:

(i)   harmonic regression: least squares regression on a sinusoid or sinusoids (Chapter 2);

(ii) harmonic analysis: the discrete Fourier transform, periodogram analysis (Chapters 3, 4, and 5);

(iii) complex demodulation (Chapter 6);

(iv) spectrum analysis (Chapters 7, 8, and 9).

The order of the discussion is dictated by the increasing complexity of the statistical concepts involved. At all stages of the book, the reader is urged to stop and consider the appropriateness of applying a particular method to the set of data under consideration. In several cases some preprocessing is carried out to make the data more appropriate. This is all designed to make the point that any data-analytic procedure based on the sine and cosine functions has a better chance of yielding useful conclusions if the data show some kind of oscillation, preferably as uniform or as regular as possible.

There are exercises at the ends of most sections. Some are algebraic manipulations designed to make the reader more familiar with the tools of discrete Fourier analysis and to build his or her confidence. The others are used to indicate some of the directions in which the theory of time series analysis has revealed useful results. However, the most important exercise, and one that should be omitted by no serious reader, is the analysis of data. Almost all the data used in this book are widely available. However the reader who tries the methods on data arising in his or her own field will gain the added benefit of seeing these data from a new point of view. Many of the more general purpose computer programs used to analyze the examples have been included in Appendices to the relevant chapters. They are coded in a moderately transportable dialect of FORTRAN (apart from the use of the symbol $\neq$ rather than $'$ to delimit character strings in FORMAT statements), and all have been executed successfully. Naturally they are not guaranteed to be bug-free.

I was encouraged to write this book by Geoffrey Watson, who saw clearly the need for an introductory text on Fourier methods not encumbered by an abundance of mathematical, probabilistic, or statistical detail. An early version was used as class notes in an undergraduate course on time series taught in the Department of Statistics at Princeton University in the spring of 1974. Revision was begun during a visit to the Computer Centre and the Department of Statistics, Institute of Advanced Study, Australian National University, Canberra. The final draft was prepared during a leave spent at the Department of Statistics, University of California at Berkeley. I am grateful to colleagues at all three institutions for their assistance and encouragement, especially David Brillinger, Richard Hamming, E. J. Hannan, and John Tukey. Each institution provided an excellent background within which to work, including library and computer facilities.

The plasma physics data used in Chapters 6 and 9 were kindly provided by Joseph Cecchi of the Plasma Physics Laboratory, Department of Astrophysics, Princeton University. I also thank Michael Stoto and John Turner for their assistance with computing problems. The final draft was typed immaculately, with great good humor, by Ruth Suzuki. Much of the computing was supported by the Office of Naval Research, under contract number 0014-67-A-0151-0017 with the Department of Statistics, Princeton University.

PETER BLOOMFIELD

Berkeley, California
May 1975

CONTENTS

1

INTRODUCTION

In its simplest form a *time series* is a collection of numerical observations arranged in a natural order. Usually each observation is associated with a particular instant of time or interval of time, and it is this that provides the ordering. The observations could equally well be associated with points along a line, but whenever they are ordered by a single variable we refer to it conventionally as "time." We shall generally assume that the time values are equally spaced.

There are many situations in which more than one phenomenon is observed at each time point, and this gives rise to *multiple time series*. In other situations each observation is associated with the values of a number of variables (or in other words with a point in a space of dimension higher than 1). Such data are called *spatial series*, the most common form being observations associated with points in the plane. Both of these types of data are generalizations of the simple time series, and each requires the appropriate extension of simple time series methods. However, most of the considerations that arise in the analysis of such series arise also in the analysis of simple time series, to which the bulk of this book is devoted.

1.1 FOURIER ANALYSIS

In its narrowest sense, the *Fourier analysis* or *harmonic analysis* of a time series is a decomposition of the series into a sum of sinusoidal components (the coefficients of which are the *discrete Fourier transform* of the series). However, we use the term in a wider sense to describe any data-analysis

1

procedure that describes or measures the fluctuations in a time series by comparing them with sinusoids.

A set of data that is used as an example in Chapters 2 and 5 is shown in Figure 1.1. It represents the magnitudes of a variable star at midnight on 600 successive nights (from Whittaker and Robinson, 1924, p. 349). We shall show that these data consist approximately of the sum of two sinusoidal components, and in this case Fourier analysis in its narrower sense (with some slight modifications) provides an accurate and economical description of the data. Figure 1.2 shows a second set of data used as an example, the Wolf sunspot numbers, as tabulated by Waldmeier (1961). Although these data contain a clear succession of peaks occurring roughly every 11 years, they are not regular enough to be represented by any one sinusoid. A local form of harmonic analysis known as *complex demodulation* may be used to describe the oscillations in these data.

Figure 1.3 shows the index of wheat prices in Western Europe presented by Beveridge (1921). Again there is a succession of peaks in the data, but no tendency for them to occur in any regular fashion is evident. Harmonic analysis reveals that there are few if any persistent sinusoidal components in these data. Nevertheless, the oscillations in this series (or a suitably transformed series) may be described in sinusoidal terms by a *spectrum analysis.* This is a method that describes the *tendency* for oscillations of a given frequency to appear in the data, rather than the oscillations themselves.

All of these methods fall under the heading of Fourier analysis in its wider sense. In the following chapters the methods are described in detail, and by applying them to various sets of data we show how they may be used to draw various inferences about the data being analyzed.

1.2 HISTORICAL DEVELOPMENT OF FOURIER METHODS

The simplest periodic data are those consisting of a single cosine wave

$$x_t = R \cos(\omega t + \phi),$$

observed with very little error. For example, the Romans had sufficiently accurate observations of the apparent motion of the sun to know that the length of the year is approximately $365\frac{1}{4}$ days, the value used to construct the Julian calendar.

Data that contain more than one periodic component are more difficult to analyze. In 1772 Lagrange proposed a method based on the use of rational functions to identify such components, and used it to analyze the orbit of a comet (see Lagrange, 1873). Like related methods proposed by

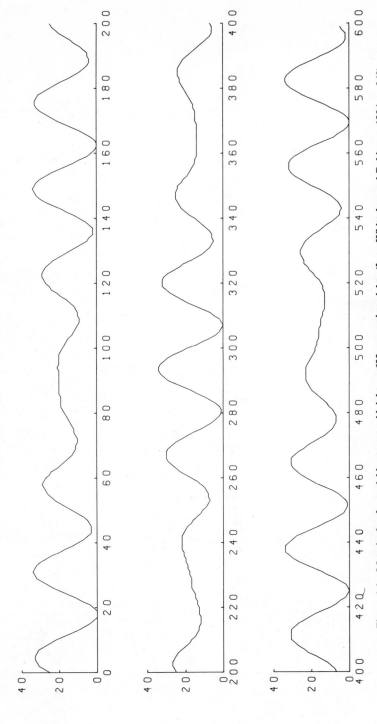

Figure 1.1 Magnitude of a variable star at midnight on 600 successive nights (from Whittaker and Robinson, 1924, p. 349).

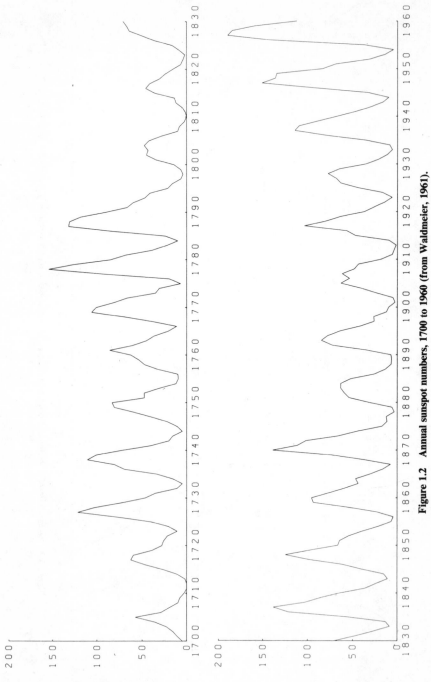

Figure 1.2 Annual sunspot numbers, 1700 to 1960 (from Waldmeier, 1961).

4

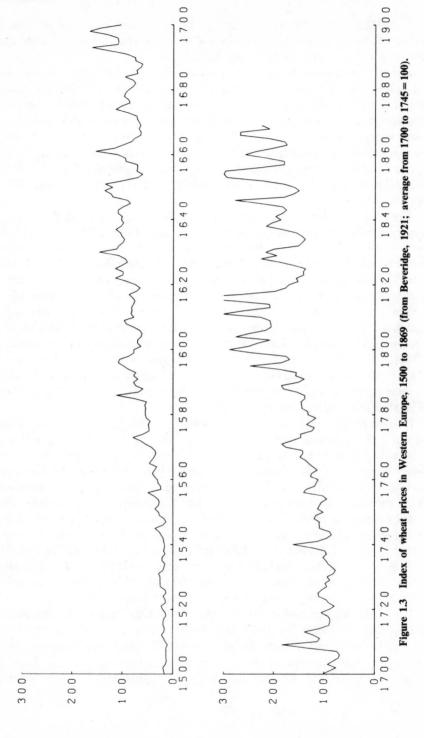

Figure 1.3 Index of wheat prices in Western Europe, 1500 to 1869 (from Beveridge, 1921; average from 1700 to 1745 = 100).

5

Dale (1914a, b) and Prony (Hamming, 1973, Section 39.4), the method is tedious to apply to any but the shortest of series, and it is very sensitive to errors or other disturbances in the data.

The first procedure to be used at all widely, and the first to be feasible for moderate numbers of data points, was described in 1847 by Buys-Ballot. This is a tabular method and is most easily used to detect a periodic component for which one cycle covers a whole number of observations. A more sophisticated version, described by Stewart and Dodgson (1879), may be used with some computational effort to improve the estimate of an approximately known period. The procedure is discussed in detail by Whittaker and Robinson (1924, Chapters 10 and 13) and Anderson (1971, pp. 106–112).

The Fourier analysis of a series of numbers may be carried out by a similar tabular technique, described by Schuster (1897). This method was used in the second half of the nineteenth century to find periodic components of known periods in tidal data, meteorological series, and astronomical series. However, the computations required were repetitive and tedious. Again the method is easiest if the period of the component covers a whole number of observations. Thomson (1876, 1878) described and built an instrument for carrying out this analysis mechanically and claimed that it would reduce the time needed for an analysis by a factor of 10. Stokes (1879), in a comment on the report of Stewart and Dodgson (1879), pointed out that Thomson's harmonic analyzer could also be used to determine the unknown period of a component in a series. The method is related to complex demodulation.

However, when Fourier analysis is used to search for periodic components of unknown period in empirical data, the results may be very misleading. For instance, Knott (1897) claimed to have discovered components with periods related to the lunar cycle in a series of Japanese earthquakes. Schuster (1897) then showed that their magnitudes were not large enough to be statistically significant. Schuster (1898) provided additional discussion of the Fourier analysis of empirical data, introduced the *periodogram*, and made many penetrating comments on its use. In some further papers (Schuster, 1900, 1906) he applied these ideas to the analysis of various sets of data, including the sunspot series (Figure 1.2). Beveridge (1921, 1922) gave an extensive periodogram analysis of the wheat-price index (Figure 1.3), which was a major computational venture.

The probabilistic and statistical theory of time series was developed during the the 1920s and 1930s (see Wold, 1954), and the concept of the *spectrum* of a series was introduced. In the 1940s and 1950s there was great interest in the problem of *estimating* the spectrum of a series. Daniell (1946) pointed out that a smoothed form of the periodogram is a suitable estimate, and this was followed up by Bartlett (1948) and Kendall (1948), while Hamming and Tukey (1949) proposed a slightly different procedure. There followed a rapid development of the theory and practice of spectrum estimation, to which major contributions were made by Grenander and Rosenblatt (1953, 1957), Parzen (1957a, b), and Blackman and Tukey (1959). The stimulus for this development was the increase in the use of Fourier methods in many fields, principally electrical engineering in its diverse forms, and the parallel increase in the availability of electronic digital computers to carry out the extensive computations involved.

The next major advance was in the computation of Fourier transforms of data. Cooley and Tukey (1965) described an algorithm that significantly reduces the computational effort involved. The fast Fourier transform, as it became known, together with advances in computer technology, has made feasible the routine Fourier analysis of extensive sets of data (see, for instance, Brigham, 1974). The spectrum estimation techniques described in Chapters 7 and 9 are designed to take full advantage of these computational advances.

1.3 WHY USE TRIGONOMETRIC FUNCTIONS?

The essence of Fourier analysis is the representation of a set of data in terms of sinusoidal functions. At this point, it seems appropriate to justify the choice of these functions, since many other families of periodic functions share at least some of the properties of the sinusoids. Any of these families could be used in a similar way, and there are situations in which special considerations make a nonsinusoidal family more suitable.

The most basic property of the sinusoids that makes them generally suitable for the analysis of time series is their simple behavior under a change of time scale. A sinusoid of *frequency* ω (in radians per unit time) or *period* $2\pi/\omega$ may be written as

$$f(t) = R\cos(\omega t + \phi),$$

where R is the *amplitude* and ϕ is the *phase*. If we change the time variable to $u = (t - a)/b$, which incorporates a change both of origin and of scale, this becomes

$$g(u) = f(a + bu)$$

$$= R\cos(\omega bu + \phi + \omega a)$$

$$= R'\cos(\omega' u + \phi'),$$

say, where $R' = R$, $\omega' = \omega b$, and $\phi' = \phi + \omega a$. Thus the amplitude is unchanged, the frequency is multiplied by b (the *reciprocal* of the change in the time scale), and the phase is altered by an amount involving the change of time origin and the frequency of the sinusoid. Since the time origin associated with a set of data is often arbitrary, these simple relationships are useful. In particular, since the amplitude of the sinusoid depends on neither the origin nor the scale of the time variable, it may be regarded as an absolute quantity with no arbitrariness in its definition.

A further useful feature of the sinusoids is their behavior under *sampling* (that is, observing a function of the continuous variable t at an equally spaced set of values $t_0, t_1, \ldots$), for if the sampling interval is Δ, the sinusoids

$$R\cos(\omega_1 t + \phi) \qquad \text{and} \qquad R\cos(\omega_2 t + \phi)$$

are indistinguishable if $\omega_1 - \omega_2$ is a multiple of $2\pi/\Delta$. This phenomenon, known as *aliasing*, is discussed further in Section 2.5.

2

THE SEARCH
FOR PERIODICITY

In this chapter we examine the problem of describing what periodicities if any are present in a given set of data. In some cases we know a collection of periods that may be expected to be present, and we have to find the associated amplitudes and phases. However, often we have no prior information about the periods, and these too must be found. The first problem is simpler, and we discuss it first. From this discussion we shall develop a way of solving the more difficult problem.

2.1 A CURVE-FITTING APPROACH

As an example consider the variable-star data of Figure 1.1. Over the 600 days of data we count 21 peaks; this suggests that any periodicity should have a period of around $600/21 \cong 28.6$ days. Thus the tth data value should contain a component of the form $R\cos(\omega t + \phi)$, where $\omega = 2\pi/28.6 = 0.220$. We *model* the data as

$$x_t = \mu + R\cos(\omega t + \phi) + \varepsilon_t, \qquad t = 0, 1, \ldots, 599, \qquad (1)$$

the simplest case of the "hidden periodicities" model. Here x_t denotes the tth data value, and ε_t is the tth *residual* (that is, whatever is needed to make the equality exact). We regard the model as good (and say that it *fits* the

9

data well) if the residuals are generally small. The term μ is an added constant. Since a cosine wave oscillates about 0, while the data oscillate between 0 and around 30, such a term is clearly needed if the residuals are to be at all small.

The unknown parameters are μ, R, and ϕ, and in the next section we show how to find values for them that make the residuals as small as possible in a certain specific sense. Initially we shall keep ω fixed at 0.220, but in Section 2.3 we shall regard it as an additional unknown and find a better value.

For the purposes of this chapter, we shall follow the common practice of measuring the size of the residuals by the sum of their squared values. Thus the problem is to find μ, R, and ϕ (and, later ω) to minimize

$$S(\mu, R, \phi) = S(\mu, R, \phi, \omega) = \sum_{t=0}^{599} \{x_t - \mu - R\cos(\omega t + \phi)\}^2,$$

the term between braces being precisely the tth residual for given values of μ, R, and ϕ (and ω). This is an example of the method of *least squares*. Least squares methods are widely used and have many computational and theoretical advantages. However, they also have certain deficiencies, which will be mentioned briefly in Section 5.3.

It is easily seen that least squares problems are simplest when the model is a linear function of the unknown parameters, since then the function to be minimized is quadratic. Equation 1 is nonlinear in R and ϕ, but may be rewritten as

$$x_t = \mu + A\cos\omega t + B\sin\omega t + \varepsilon_t,$$

where $A = R\cos\phi$ and $B = -R\sin\phi$. Furthermore, given any values of A and B, we may solve for R and ϕ. We may therefore regard A and B as the parameters, and the model is now linear, for fixed ω. In the next section we solve the elementary problem of finding μ, A, and B for fixed ω. In Section 2.3 we show how our current estimate of the frequency ω may be improved.

Exercise 2.1 Least Squares Straight Line

Suppose that $(x_1, y_1), \ldots, (x_n, y_n)$ are a set of points in the plane. It is sometimes useful to model such a set of points by a straight line, $y = a + bx$. The *least squares straight line* has parameters $\hat{a}$ and $\hat{b}$ which minimize the sum of squared residuals,

$$S(a,b) = \sum_{i=1}^{n} (y_i - a - bx_i)^2.$$

(i)   Verify that, provided the x-values are not all the same,

$$\hat{b} = \frac{\sum\limits_{i=1}^{n} y_i(x_i - \bar{x})}{\sum\limits_{i=1}^{n} (x_i - \bar{x})^2}$$

and

$$\hat{a} = \bar{y} - \hat{b}\bar{x},$$

where $\bar{x} = (x_1 + x_2 + \cdots + x_n)/n$, and $\bar{y}$ is similarly defined.

(ii)   Find the corresponding formulas for the coefficients of the least squares parabola, $y = a + bx + cx^2$.

2.2   LEAST SQUARES ESTIMATION OF AMPLITUDE AND PHASE

In this section we show how to estimate the parameters of a sinusoid, with or without an added constant. The frequency ω is regarded as known and is not varied to improve the fit. In the next section the method is extended to include the estimation of ω. We consider first the simple two-parameter model of a sinusoid with no added constant. The model is

$$x_t = A \cos \omega t + B \sin \omega t + \varepsilon_t,$$

and the principle of least squares leads us to minimize

$$T(A, B) = \sum_{t=0}^{n-1} (x_t - \mu - A \cos \omega t - B \sin \omega t)^2,$$

restricting μ to be zero for the present, and keeping ω fixed. Now

$$\frac{\partial T}{\partial A} = -2 \sum \cos \omega t (x_t - A \cos \omega t - B \sin \omega t),$$

$$\frac{\partial T}{\partial B} = -2 \sum \sin \omega t (x_t - A \cos \omega t - B \sin \omega t),$$

and the equations that result from equating these to zero have the solution

$$A = \hat{A} = \frac{1}{\Delta} \left\{ \sum x_t \cos \omega t \sum (\sin \omega t)^2 \right.$$

$$\left. - \sum x_t \sin \omega t \sum \cos \omega t \sin \omega t \right\},$$

$$B = \hat{B} = \frac{1}{\Delta} \left\{ \sum x_t \sin \omega t \sum (\cos \omega t)^2 \right. \tag{2}$$

$$\left. - \sum x_t \cos \omega t \sum \cos \omega t \sin \omega t \right\},$$

where

$$\Delta = \sum (\cos \omega t)^2 \sum (\sin \omega t)^2 - \left(\sum \cos \omega t \sin \omega t \right)^2.$$

The sums involving only trigonometric functions may be evaluated, using the results of Exercise 2.2, to give

$$\sum (\cos \omega t)^2 = \frac{n}{2} \left\{ 1 + D_n(2\omega) \cos(n-1)\omega \right\},$$

$$\sum \cos \omega t \sin \omega t = \frac{n}{2} D_n(2\omega) \sin(n-1)\omega,$$

$$\sum (\sin \omega t)^2 = \frac{n}{2} \left\{ 1 - D_n(2\omega) \cos(n-1)\omega \right\},$$

where

$$D_n(\omega) = \frac{\sin n\omega/2}{n \sin \omega/2}$$

is a version of the *Dirichlet kernel* (Titchmarsh, 1939, p. 402). The sums involving $\{x_t\}$ usually have to be evaluated directly.

To find R and ϕ, the amplitude and phase, we solve the equations $A = R\cos\phi$ and $B = -R\sin\phi$. Since R is nonnegative, it follows that $R = (A^2 + B^2)^{1/2}$. The basic equation for ϕ is $\tan\phi = -B/A$. However, the solution $\phi = \arctan -B/A$ is incorrect, since this gives the same value for $-A$ and $-B$ as for A and B. The full solution is as follows:

$$\phi = \begin{cases} \arctan(-B/A), & A > 0, \\ \arctan(-B/A) - \pi, & A < 0, B > 0, \\ \arctan(-B/A) + \pi, & A < 0, B \leqslant 0, \\ -\pi/2, & A = 0, B > 0, \\ \pi/2, & A = 0, B < 0, \\ \text{arbitrary}, & A = 0, B = 0. \end{cases}$$

(The FORTRAN function ATAN2($-B, A$) returns the required value.)

The model that seems appropriate for the variable-star data is the three-parameter "sinusoid plus constant" model given in Section 2.1,

$$x_t = \mu + A \cos \omega t + B \sin \omega t + \varepsilon_t.$$

The equations for the least squares estimates of μ, A, and B (which we shall denote as $\hat{\mu}$, $\hat{A}$, and $\hat{B}$, respectively) are

$$\sum (x_t - \mu - A \cos \omega t - B \sin \omega t) = 0,$$

$$\sum \cos \omega t (x_t - \mu - A \cos \omega t - B \sin \omega t) = 0, \tag{3}$$

$$\sum \sin \omega t (x_t - \mu - A \cos \omega t - B \sin \omega t) = 0.$$

These too may be solved explicitly (see Exercise 2.3). For the variable-star data with $\omega = 0.220$, we find

$$\hat{\mu} = 17.11, \qquad \hat{A} = -1.102, \qquad \hat{B} = 8.406,$$

and hence the estimates of R and ϕ are

$$\hat{R} = 8.478, \qquad \hat{\phi} = -1.701.$$

Note that for negative $\hat{\phi}$ the argument $\omega t + \hat{\phi}$ of $\cos(\omega t + \hat{\phi})$ first vanishes at $t = |\hat{\phi}|/\omega \approx 7.7$. Thus the fitted cosine wave has a peak at $t = 7.7$, while the first peak in the data is between $t = 4$ and $t = 5$. Since the peaks in the data are not evenly spaced, this seems to be reasonable agreement.

We can find very useful approximations to (3) as follows (see Exercise 2.4). When the purely trigonometric sums are evaluated, the coefficients in (3) involve the term $n/2$ and terms such as $(n/2)D_n(\omega/2)$. We note first that $D_n(2k\pi/n) = 0$ for any integer k, and that $|nD_n(\omega)| \leqslant 1/(\sin \omega/2)$. This means that the terms in (3) involving D_n are, for large n and ω not too close to 0, always small compared with $n/2$, and sometimes exactly 0. By omitting all these terms we find the reduced set of equations

$$n\mu = \sum x_t,$$

$$\frac{nA}{2} = \sum x_t \cos \omega t,$$

$$\frac{nB}{2} = \sum x_t \sin \omega t.$$

(We note in passing that the second and third equations are also approximations to the two-parameter equations (2).)

It often happens that μ is larger than A or B, and in this case it is unwise to ignore any term involving μ. The approximations are then

$$n\mu = \sum x_t,$$

$$\mu \sum \cos \omega t + \frac{nA}{2} = \sum x_t \cos \omega t, \qquad (4)$$

$$\mu \sum \sin \omega t + \frac{nB}{2} = \sum x_t \sin \omega t.$$

The solutions to these equations (denoted as $\tilde{\mu}, \tilde{A}, \tilde{B}$) are

$$\tilde{\mu} = \bar{x} = \frac{1}{n} \sum x_t,$$

$$\tilde{A} = \frac{2}{n} \sum (x_t - \bar{x}) \cos \omega t, \qquad (5)$$

$$\tilde{B} = \frac{2}{n} \sum (x_t - \bar{x}) \sin \omega t,$$

and may be regarded as approximate solutions to (3) (see Exercise 2.4). For the variable-star data we find $\tilde{\mu} = \bar{x} = 17.11$, $\tilde{A} = 1.103$, $\tilde{B} = 8.403$, $\tilde{R} = 8.475$, and $\tilde{\phi} = -1.701$. Notice that the differences between these values and the exact least squares values appear only in the fourth significant figure, if at all.

The adequacy of the model as a representation of the data may be assessed by examining the sum of squares of the residuals. The values of $T(\hat{\mu}, \hat{A}, \hat{B})$ and $T(\bar{x}, \tilde{A}, \tilde{B})$ are both 26,769.5, to six significant figures. The approximate solutions are very satisfactory, in that the sum of squares is increased by less than one part in a million. Either value may be compared with $T(\bar{x}, 0, 0) = 48,324.3$, the sum of squares of the residuals just from a constant term. The difference $T(\bar{x}, 0, 0) - T(\hat{\mu}, \hat{A}, \hat{B}) = 21,554.8$ [or $T(\bar{x}, 0, 0) - T(\bar{x}, \tilde{A}, \tilde{B})$] may be regarded as the amount of squared variation in the data that can be accounted for by the frequency ω. A set of approximations similar to those used to obtain (5) shows that both quantities are approximately

$$\frac{n}{2}(\tilde{A}^2 + \tilde{B}^2) = \frac{n}{2}\tilde{R}^2 = 21,546.8.$$

(See Exercise 2.5.) The error in using this approximation is therefore around 3 parts in 10,000, in the present case.

Exercise 2.2   Some Trigonometric Identities

(i)   Show that

$$\sum_{t=0}^{n-1} \exp(i\lambda t) = \frac{\exp(in\lambda) - 1}{\exp(i\lambda) - 1}$$

$$= \exp\left\{ \frac{i(n-1)\lambda}{2} \right\} \frac{\exp(in\lambda/2) - \exp(-in\lambda/2)}{\exp(i\lambda/2) - \exp(-i\lambda/2)}.$$

(ii)   Use the Euler relation

$$\exp(i\lambda) = \cos\lambda + i\sin\lambda$$

and its inverse

$$\cos\lambda = \tfrac{1}{2}\{\exp(i\lambda) + \exp(-i\lambda)\}, \qquad \sin\lambda = \frac{1}{2i}\{\exp(i\lambda) - \exp(-i\lambda)\}$$

to deduce that

$$\sum \cos\lambda t = \cos\left\{ \frac{(n-1)\lambda}{2} \right\} \frac{\sin n\lambda/2}{\sin\lambda/2},$$

$$\sum \sin\lambda t = \sin\left\{ \frac{(n-1)\lambda}{2} \right\} \frac{\sin n\lambda/2}{\sin\lambda/2}.$$

(iii)   Use the addition formulas

$$\sin(\lambda + \mu) = \sin\lambda\cos\mu + \cos\lambda\sin\mu,$$

$$\cos(\lambda + \mu) = \cos\lambda\cos\mu - \sin\lambda\sin\mu,$$

and their inverses

$$\cos\lambda\cos\mu = \tfrac{1}{2}\{\cos(\lambda + \mu) + \cos(\lambda - \mu)\},$$

$$\cos\lambda\sin\mu = \tfrac{1}{2}\{\sin(\lambda + \mu) - \sin(\lambda - \mu)\},$$

$$\sin\lambda\sin\mu = \tfrac{1}{2}\{\cos(\lambda - \mu) - \cos(\lambda + \mu)\},$$

to evaluate $\sum \cos\lambda t \cos\mu t$, $\sum \cos\lambda t \sin\mu t$, and $\sum \sin\lambda t \sin\mu t$. Note the special cases $\lambda = \mu$.

Exercise 2.3   Equations for the Three-Parameter "Sinusoid Plus Constant" Model

The derivatives of

$$T(\mu, A, B) = \sum_{t=0}^{n-1} (x_t - \mu - A \cos \omega t - B \sin \omega t)^2$$

with respect to μ, A, and B are

$$\frac{\partial T}{\partial \mu} = -2 \sum (x_t - \mu - A \cos \omega t - B \sin \omega t),$$

$$\frac{\partial T}{\partial A} = -2 \sum \cos \omega t (x_t - \mu - A \cos \omega t - B \sin \omega t),$$

$$\frac{\partial T}{\partial B} = -2 \sum \sin \omega t (x_t - \mu - A \cos \omega t - B \sin \omega t),$$

respectively. Simplify these expressions using the results of Exercise 2.2, and solve them for the least squares estimates $\hat{\mu}$, $\hat{A}$, and $\hat{B}$ of μ, A, and B, respectively.

Exercise 2.4   The Approximate Least Squares Estimates

(i)   For the two-parameter model (equations 2), show that

$$|\hat{A} - \frac{2}{n} \sum x_t \cos \omega t| = |D_n(2\omega)| |\hat{A} \cos(n-1)\omega + \hat{B} \sin(n-1)\omega|$$

$$\leqslant \frac{\hat{R}}{n \sin \omega},$$

and that similarly

$$|\hat{B} - \frac{2}{n} \sum x_t \sin \omega t| \leqslant \frac{\hat{R}}{n \sin \omega}.$$

(ii)   For the three-parameter model (equations 3), show that

$$|\hat{\mu} - \bar{x}| \leqslant \frac{\hat{R}}{n \sin \omega/2}$$

and that both

$$|\hat{A} - \frac{2}{n} \sum x_t \cos \omega t| \qquad \text{and} \qquad |\hat{B} - \frac{2}{n} \sum x_t \sin \omega t|$$

are bounded by

$$\frac{2\hat{\mu}}{n\sin\omega/2} + \frac{\hat{R}}{n\sin\omega}.$$

(iii)   For the three-parameter model show that both $|\hat{A} - \tilde{A}|$ and $|\hat{B} - \tilde{B}|$ are bounded by

$$\hat{R}\left\{\frac{2}{(n\sin\omega/2)^2} + \frac{1}{n\sin\omega}\right\}.$$

Exercise 2.5   The Sum of Squares Associated with ω

(i)   For the two-parameter model, show that

$$T(0,\hat{A},\hat{B}) = \sum x_t^2 - \left\{\left(\sum y_t\cos\omega t\right)^2 \sum(\sin\omega t)^2\right.$$

$$-2\sum y_t\cos\omega t\sum y_t\sin\omega t\sum\cos\omega t\sin\omega t$$

$$\left.+\left(\sum y_t\sin\omega t\right)^2\sum(\cos\omega t)^2\right\}/\Delta.$$

NOTE: This may be interpreted as

sum of squares of residuals

= sum of squares of original data

− sum of squares associated with frequency ω.

(ii)   Find the corresponding expression for the residual sum of squares $T(\hat{\mu},\hat{A},\hat{B})$ in the three-parameter model.

(iii)   Show that the sum of squares associated with ω is approximately

$$\frac{2}{n}\left\{\left(\sum x_t\cos\omega t\right)^2 + \left(\sum x_t\sin\omega t\right)^2\right\}$$

in the two-parameter model, and

$$\frac{2}{n}\left[\left\{\sum(x_t-\bar{x})\cos\omega t\right\}^2 + \left\{\sum(y_t-\bar{x})\sin\omega t\right\}^2\right]$$

$$= \frac{n}{2}(\tilde{A}^2 + \tilde{B}^2) = \frac{n}{2}\tilde{R}^2$$

in the three-parameter model.

2.3 LEAST SQUARES ESTIMATION OF FREQUENCY

In this section we show how the methods of Section 2.2 may be extended to include the estimation of the frequency ω. We shall deal only with the three-parameter "sinusoid plus constant" model, since that is the more generally useful. However, an exactly analogous method could be used to estimate ω in the two-parameter case. In the next section we shall extend the method further to the fitting of a number of frequencies (in fact, using the two-parameter method as the basic building block).

In Section 2.2 we found values $\hat{\mu}$, $\hat{A}$, and $\hat{B}$ of μ, A, and B, respectively, to minimize

$$T(\mu, A, B) = T(\mu, A, B, \omega)$$

$$= \sum_{t=0}^{n-1} (x_t - \mu - A\cos\omega t - B\sin\omega t)^2,$$

for a fixed ω. It was shown that these are approximately

$$\tilde{\mu} = \bar{x} = \frac{1}{n}(x_0 + \cdots + x_{n-1}),$$

$$\tilde{A}(\omega) = \frac{2}{n}\sum(x_t - \bar{x})\cos\omega t,$$

$$\tilde{B}(\omega) = \frac{2}{n}\sum(x_t - \bar{x})\sin\omega t,$$

and furthermore that the residual sum of squares is

$$T\{\hat{\mu}(\omega), \hat{A}(\omega), \hat{B}(\omega), \omega\} \cong T\{\bar{x}, \tilde{A}(\omega), \tilde{B}(\omega), \omega\}$$

$$\cong T(\bar{x}, 0, 0, \omega) - \frac{n}{2}\{\tilde{A}(\omega)^2 + \tilde{B}(\omega)^2\}$$

$$= T(\bar{x}, 0, 0, \omega) - \frac{n}{2}\tilde{R}(\omega)^2,$$

where $\tilde{R}(\omega)^2 = \tilde{A}(\omega)^2 + \tilde{B}(\omega)^2$. In this section ω is regarded as an additional unknown, and so the dependence of the various estimates on ω is shown explicitly. The best value for ω in the sense of least squares is the value $\hat{\omega}$, that minimizes $T\{\hat{\mu}(\omega), \hat{A}(\omega), \hat{B}(\omega), \omega\}$. The corresponding approximation is the value $\tilde{\omega}$ that maximizes $\tilde{R}(\omega)^2$. Figure 2.1 shows a plot of part of the latter function for the variable-star data (Figure 1.1). An equivalent function used in later chapters is the *periodogram*

$$I(\omega) = \frac{n}{8\pi}\tilde{R}(\omega)^2.$$

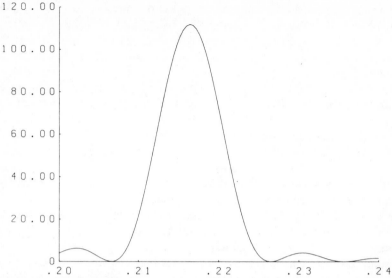

Figure 2.1   Periodogram of the variable-star data for frequencies ω, $0.20 < \omega < 0.24$.

Often, as in searching for a maximum or comparing function values, the actual values of the function are unimportant, and such rescalings have no impact. We shall therefore also refer to $\tilde{R}(\omega)^2$ as the periodogram when the difference is irrelevant.

The function

$$\frac{2}{n}\left[T(\bar{x},0,0,0) - T\{\hat{\mu}(\omega),\hat{A}(\omega),\hat{B}(\omega),\omega\}\right]$$

was also calculated. However, it differed from $\tilde{R}(\omega)^2$ by at most 0.8, and hence it was not graphed. The graph shows a clear maximum at a value somewhat less than 0.220. The actual peak was found to be at $\omega=0.21644$. The peak of the ungraphed function above was found to be at $\omega=0.21641$. The difference is minor, especially since it is smaller than the difference between either value and the value found in Section 2.4. By contrast, the difference *is* appreciable in the light of the statistical results described in Section 2.6.

The least squares estimates of the parameters of a sinusoid with frequency $\omega=0.21644$ are

$$\hat{\mu}=17.08,\qquad \hat{A}=8.480,\quad \hat{B}=6.211,$$
$$\tilde{\mu}=\bar{x}=17.11,\quad \tilde{A}=8.550,\quad \tilde{B}=6.225.$$

The residual sum of squares for $\hat{\mu}$, $\hat{A}$, and $\hat{B}$ is 14,977.2, while that for $\bar{x}$, $\tilde{A}$, and $\tilde{B}$ is greater by 2.0, a negligible amount.

The graph also shows a subsidiary peak, or *sidelobe*, occurring on either side of the main peak, and separated from it by a trough in which the value is indistinguishable from zero. We shall see in Chapter 3 that this is typical of such graphs. The sidelobes do not indicate the presence of other periodicities.

The maxima of these functions were found numerically, using an algorithm described by Brent (1972). The derivatives of both functions with respect to ω are highly nonlinear and have many zeros (indeed, Figure 2.1 shows that there are six zeros just in the interval $0.20 \leqslant \omega \leqslant 0.24$). This makes an analytic solution impossible and also renders numerical methods based on the gradient treacherous. For instance, Newton's method could easily lead us to one of the other stationary points. (The FORTRAN program used is presented in the Appendix to this chapter.)

Figure 2.2 shows the residuals, that is, the original data less the fitted cosine term. They have a very pronounced periodicity with a period of around 24 days, or a frequency of approximately $\omega = 0.262$. Thus the original data must have contained at least these two periodic components. The estimation of a number of frequencies is described in the next section; in particular, we shall show that the presence of a second periodic component, especially one with a similar frequency, can noticeably distort the estimates of frequency and of amplitude and phase.

2.4 MULTIPLE PERIODICITIES

It emerged at the end of the preceding section that the data being used as an example actually contain more than one periodic component. In this section we describe how a number of components may be estimated, again using the variable-star data as an example.

The simplest procedure would be to repeat the analysis of Section 2.3, but searching now for a maximum near $\omega = 0.262$, the second frequency. If we distinguish quantities associated with the first (or second) frequency by the subscript 1 (or 2), we find that $\tilde{\omega}_2 = 0.2621$, $\tilde{A}_2 = -0.7579$, and $\tilde{B}_2 = 7.727$. However, the expressions for these estimates were obtained by the least squares fitting of the model

$$x_t = \mu + A \cos \omega t + B \sin \omega t + \varepsilon_t,$$

where ε_t is the tth residual. The idea behind a least squares method is to make these residuals as small as possible. In the present case, however, the residual term necessarily includes the strong periodic component found in the preceding section, and it does not make sense to try to make this small.

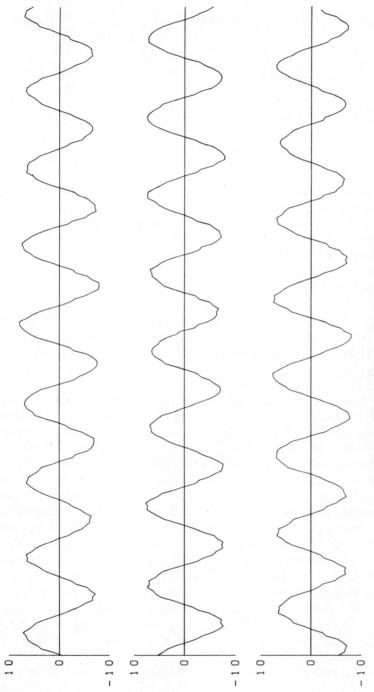

Figure 2.2   Variable-star data with fitted sinusoid subtracted.

A better approach is to include the second component in the model:

$$x_t = \mu + A_1 \cos\omega_1 t + B_1 \sin\omega_1 t + A_2 \cos\omega_2 t + B_2 \sin\omega_2 t + \varepsilon_t.$$

This leads us to minimize

$$\sum_{t=0}^{n-1} (x_t - \mu - A_1 \cos\omega_1 t - B_1 \sin\omega_1 t - A_2 \cos\omega_2 t - B_2 \sin\omega_2 t)^2$$

$$= T(\mu, A_1, B_1, \omega_1, A_2, B_2, \omega_2), \qquad \text{say.} \qquad (6)$$

The most natural extension of the method used in the Section 2.3 to find a single frequency is as follows. First, we note that for fixed ω_1 and ω_2 the model is linear in the remaining parameters. Hence the conditionally best values of these may be found by conventional methods and substituted in the function T (see Exercise 2.6). This gives us a new function

$$U(\omega_1, \omega_2) = T(\hat{\mu}, \hat{A}_1, \hat{B}_1, \omega_1, \hat{A}_2, \hat{B}_2, \omega_2),$$

where $\hat{\mu}$, $\hat{A}_1$, $\hat{B}_1$, and $\hat{B}_2$ are all functions of both ω_1 and ω_2. The function U or an appropriate approximation could then be minimized numerically by one of the methods generally available (see Brent, 1972). Note that, by analogy with the functions examined in the preceding section, we can expect U to have many stationary points.

An alternative approach, which also builds on the method of Section 2.3 but avoids the explicit two-dimensional search, is based on the method of *cyclic descent*. The general idea of a cyclic descent method is to divide the parameters into subsets (exhaustive and usually exclusive), in such a way that the optimization with respect to parameters in any one subset, holding the remaining parameters fixed, can be done fairly easily. The method then is to update the subsets successively by solving these manageable optimization problems in turn. The basic method cycles through the subsets in some fixed order, until a complete cycle results in an effectively zero change in the function to be optimized. In sophisticated algorithms the subsets may be chosen in a different sequence so as to accelerate the convergence of the method, but we shall not do this. When the function cannot be reduced by varying any of the subsets of parameters, a (local) minimum has usually been reached. (For functions with continuous partial derivatives this is always the case, except for some pathological examples. However, the method can easily fail with functions that have discontinuous partial derivatives.)

For the sake of generality we shall describe a procedure using cyclic descent to fit the more general model

$$x_t = \mu + \sum_{j=1}^{m} (A_j \cos \omega_j t + B_j \sin \omega_j t) + \varepsilon_t,$$

the model of "hidden periodicities," by least squares. First, minimization with respect to μ for fixed values of the other parameters is straightforward. The optimal value is just the mean of the "corrected" series

$$x_t - \sum_{j=1}^{m} (A_j \cos \omega_j t + B_j \sin \omega_j t), \qquad t = 0, \ldots, n-1.$$

Next, if we vary ω_k, A_k, and B_k and hold the other parameters fixed, the problem is to minimize

$$\sum_{t=0}^{n-1} \left\{ x_t - \mu - \sum_{\substack{j=1 \\ j \neq k}}^{m} (A_j \cos \omega_j t + B_j \sin \omega_j t) - A_k \cos \omega_k t - B_k \sin \omega_k t \right\}^2$$

$$= \sum_{t=0}^{n-1} (y_t - A_k \cos \omega_k t - B_k \sin \omega_k t)^2,$$

where

$$y_t = x_t - \mu - \sum_{j \neq k} (A_j \cos \omega_j t + B_j \sin \omega_j t).$$

The optimization with respect to ω_k, A_k, and B_k may be done as in Section 2.3, with the simplification that the single-frequency model does not include the added constant term.

Thus one cycle of the method consists of these two steps:

(i) correct the data for all periodic components and estimate μ by the mean of the corrected series;

(ii) for k running from 1 to m, correct the series for the mean μ and the other components; then estimate ω_k, A_k, and B_k from the corrected series.

A FORTRAN program based on this algorithm is presented in the Appendix to this chapter. It contains a switch (the logical variable AP-PFLG) which allows one to select exact or approximate least squares for the single-frequency optimizations.

The results for the variable-star data are given in Table 2.1, together with the values found by the single-frequency method applied to the original data.

Note that the single-frequency estimates of frequency are very close to the values found by the cyclic descent method of this section. The sine and cosine coefficients are also similar, although not to the same extent. However, the residual sums of squares show that the cyclic descent estimates in fact provide a noticeably better fit to the data. Our conclusion is that, in the presence of strong components such as these, the single-frequency estimates do not give an adequate approximation to the least squares problem and are not satisfactory as estimates of the respective parameters. We could describe this by saying that the components *interfere* with each other. The interference is as strong as it is only because both components are strong and their frequencies are similar. A heuristic motivation for the cyclic descent method is that at each stage we remove all components other than the one currently being fitted, and thus avoid such interference.

Table 2.1   Different parameter estimates for the two-component model

	Residual Sum of Squares	Component	Frequency	A	B
Method I[a]	276.0	1	0.21644	8.5495	6.2108
		2	0.26211	-0.7579	7.7269
Method II[b]	59.6	1	0.21669	7.6551	6.5912
		2	0.26172	0.1565	7.0828
Method III[c]	54.7	1	0.21666	7.6478	6.4905
		2	0.26180	0.0006	7.0845

[a]Estimating the two components separately from the original data.

[b]Cyclic descent method, approximate least squares.

[c]Cyclic descent method, exact least squares.

The reduction in the residual sum of squares to 54.7 is remarkable. The fact that the data were reported as integers means that they contain errors at least as large as that caused by rounding off a number to the nearest integer. Since the error incurred by such rounding off is roughly uniformly distributed from $-\frac{1}{2}$ to $\frac{1}{2}$, the mean squared error would be $\frac{1}{12}$. Thus from this cause alone we could expect a residual sum of squares of around

$600 \times \frac{1}{12} = 50$. Hence the unrounded data must have been almost exactly the sum of two pure sinusoids. A further curious feature of these data is that frequencies of 0.21669 and 0.26172 correspond to periods of 28.996 and 24.008 days, respectively, which are surprisingly close to integer values for the periods of a variable star.

Exercise 2.6 Least Squares Fitting with Fixed Frequencies

(i) The equations that result from equating the partial derivatives of (6) to zero are

$$-2 \sum_{t=0}^{n-1} (x_t - \mu - A_1 \cos \omega_1 t - B_1 \sin \omega_1 t$$

$$- A_2 \cos \omega_2 t - B_2 \sin \omega_2 t) = 0,$$

$$-2 \sum_{t=0}^{n-1} \cos \omega_j t (x_t - \mu - A_1 \cos \omega_1 t - B_1 \sin \omega_1 t$$

$$- A_2 \cos \omega_2 t - B_2 \sin \omega_2 t) = 0,$$

$$-2 \sum_{t=0}^{n-1} \sin \omega_j t (x_t - \mu - A_1 \cos \omega_1 t - B_1 \sin \omega_1 t$$

$$- A_2 \cos \omega_2 t - B_2 \sin \omega_2 t) = 0,$$

$j = 1, 2$. Simplify these equations using the identities of Exercise 2.2, and show that they become diagonal if both frequencies are multiples of $2\pi/n$.

(ii) Show that if certain terms are ignored the last three equations become the same as those found in Section 2.2 for estimating the parameters of a single component of frequency ω, with $\omega = \omega_1$ and $\omega = \omega_2$, respectively. Obtain bounds for the errors introduced into the solution by this approximation.

(iii) Find bounds for the errors in the approximate solutions

$$\tilde{\mu} = \bar{x},$$

$$\tilde{A}_j = \frac{2}{n} \sum_{t=0}^{n-1} x_t \cos \omega_j t, \cdot$$

$$\tilde{B} = \frac{2}{n} \sum_{t=0}^{n-1} x_t \sin \omega_j t, \quad j = 1, 2.$$

2.5   THE EFFECT OF DISCRETE TIME: ALIASING

So far we have not discussed any restrictions that might need to be imposed on the frequency, ω, of the sinusoids being fitted to our data. Since the units of frequency are radians per unit time, it is natural to require that they be nonnegative. This may be justified by arguing that, since $\cos(-x) = \cos x$ and $\sin(-x) = -\sin x$, any cosine wave with negative frequency $-\omega$ can be written

$$A\cos(-\omega t) + B\sin(-\omega t) = A\cos\omega t + (-B)\sin\omega t,$$

as a cosine wave with a positive frequency. Thus the frequencies ω and $-\omega$ are indistinguishable; they are said to be *aliases* of each other.

The equal spacing in time of our observations introduces a further aliasing. Suppose that the *sampling interval* is Δ, so that the tth observation is made at time $t\Delta$. If the data consist of a pure cosine wave at frequency ω (for the sake of argument, with unit amplitude and zero phase), the tth observation will be

$$x_t = \cos\omega t\Delta.$$

If we increase ω from zero, this wave oscillates more and more rapidly until at $\omega = \pi/\Delta$ we have

$$x_t = \cos t\pi = (-1)^t,$$

which is clearly the most rapid oscillation we can observe. Suppose that we increase ω further, say to a value satisfying $\pi/\Delta < \omega < 2\pi/\Delta$. Let $\omega' = 2\pi/\Delta - \omega$. Then

$$x_t = \cos\omega t\Delta$$

$$= \cos\left(\frac{2\pi}{\Delta} - \omega'\right)t\Delta$$

$$= \cos(2\pi t - \omega' t\Delta)$$

$$= \cos\omega' t\Delta.$$

In the same way $\sin\omega t\Delta = -\sin\omega' t\Delta$. Thus the frequencies ω and ω' are also indistinguishable and hence are aliases of each other. We may extend the argument to any positive frequency, no matter how large.

We conclude that every frequency not in the range $0 \leqslant \omega \leqslant \pi/\Delta$ has an alias in that range, termed its *principal alias*. To avoid indeterminacy, we shall restrict frequencies to this range. Figure 2.3 shows a number of

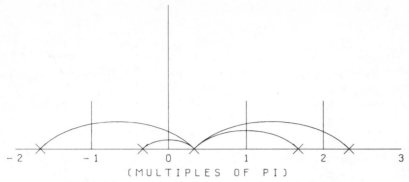

-2   -1   0   1   2   3

(M U L T I P L E S   O F   P I)

Figure 2.3   Some frequencies with the same principal alias.

frequencies with the same principal alias. The frequency π/Δ is known as the *Nyquist frequency*. It is also called the *folding frequency*, since effectively higher frequencies are folded down into the interval $[0, \pi/\Delta]$.

The Nyquist frequency is π/Δ in units of radians per unit time. In terms of cycles per unit time, it is therefore $1/(2\Delta)$. Since the sampling interval is Δ, the *sampling rate* is $1/\Delta$ observations per unit time. Thus the Nyquist frequency is one-half the sampling rate; in other words there are two samples per cycle of the Nyquist frequency, the highest frequency that can be observed.

The phenomenon of aliasing is important not only in the choice of frequencies to be fitted to data. It also must be borne in mind when designing a scheme to observe a time series. Suppose that $x(u)$ is a function of the continuous time parameter u, and that we wish to sample $x(u)$ to obtain information about frequencies in some interval, say (ω_0, ω_1). Then usually we will want the Nyquist frequency to be greater than ω_1 so that all such frequencies are directly observable. However, if $x(u)$ contains oscillations with frequencies greater than ω, we should choose the sampling frequency so that these are not aliased into the interval of interest. In fact, it is preferable when possible to remove these frequencies from the function before sampling, so that this problem cannot arise.

It should be noted that aliasing is a relatively simple phenomenon. In general, when one takes a discrete sequence of observations on a continuous function, information is lost. It is an advantage of the trigonometric functions that this loss of information is manifest in the easily understood form of aliasing.

In the chapters to follow, we shall often adopt the sampling interval as the unit of time. Then $\Delta = 1$, and the Nyquist frequency is simply π. Except where otherwise stated, this convention will be implicit.

2.6   SOME STATISTICAL RESULTS

In this chapter we have described how to obtain estimates of the coefficients of one or more sinusoidal components in a series. With the added assumption that the errors in the series are statistical or random in nature, we may describe the accuracy of those estimates.

Suppose that the data $x_0, \ldots, x_{n-1}$ were generated by the model

$$x_t = \mu + A \cos \omega t + B \sin \omega t + a_t, \tag{7}$$

where a_t are random errors or disturbances, and satisfy[†]

$$E(a_t) = 0,$$

$$E(a_t a_{t'}) = \begin{cases} v, & t = t' \\ 0, & \text{otherwise.} \end{cases}$$

The assumption that the errors at different times are uncorrelated is restrictive and is often violated in practice. We shall see in a later chapter how to make a more realistic assumption.

Having made statistical assumptions about the nature of the data, we may now make some statistical statements about the estimates discussed above. The exact least squares estimates $\hat{A}$ and $\hat{B}$ will not be considered. For the estimates $\bar{x}$, $\tilde{A}$, and $\tilde{B}$ (of μ, A, and B, respectively) we can find exact expressions for the means, variances, and covariances. These are all lengthy expressions. However, it may be shown (see Exercises 2.7 to 2.9) that

$$E(\tilde{A}) \cong A, \qquad E(\tilde{B}) \cong B, \qquad E(\bar{x}) \cong \mu,$$

$$\operatorname{var} \tilde{A} \cong \operatorname{var} \tilde{B} \cong \frac{2v}{n}, \qquad \operatorname{var} \bar{x} \cong \frac{v}{n},$$

$$\operatorname{corr}(\tilde{A}, \tilde{B}) \cong \operatorname{corr}(\tilde{B}, \bar{x}) \cong \operatorname{corr}(\bar{x}, \tilde{A}) \cong 0.$$

If we make the additional assumption that the errors $a_0, \ldots, a_{n-1}$ are independent, then by the central limit theorem (see, for instance, Feller, 1968, pp. 244,254) we would expect $\tilde{A}$, $\tilde{B}$, and $\bar{x}$, as linear functions of the a's, to be approximately normally distributed, with the stated means and variances. This may be verified by showing that the sequences of coefficients satisfy the relevant requirements.

The case in which ω is unknown and has to be estimated is more

[†]We use the conventional notation E to denote expectation.

difficult. It was first studied by Whittle (1952) and later by Walker (1971). The principal results for the estimate $\tilde{\omega}$ are that

$$E(\tilde{\omega}) = \omega + \text{terms involving } \frac{1}{n},$$

and

$$\text{var } \tilde{\omega} = \frac{24v}{n^3(A^2 + B^2)} + \text{smaller terms.}$$

At first sight, the n^{-3} behavior of var $\tilde{\omega}$ is surprising, since the variance of an estimated parameter usually behaves like the variances of $\tilde{A}$ and $\tilde{B}$, that is, like n^{-1}. However, we may easily demonstrate that a higher power is appropriate.

Consider the case in which $R^2 = A^2 + B^2$ is large compared with v. Thus the data consist of clear oscillations, with small errors superimposed. Then we can count the number of cycles in our n data points accurately, and the only uncertainty involves the magnitude of the odd fraction of a period at each end of the data. If, for instance, we see m complete cycles, but not $m + 1$, we can say that the period $2\pi/\omega$ lies between $n/(m+1)$ and n/m, or $2\pi m/n \leqslant \omega \leqslant 2\pi(m+1)/n$. Thus any estimate of ω should lie within $2\pi/n$ of the true value, and this implies that its variance is of order $1/n^2$ or better. The extra power of n is achieved by the relatively sophisticated estimate $\tilde{\omega}$.

The variance v of the errors is their mean square value. The corresponding quantity for the signal is

$$\text{ave}(A\cos\omega t + B\sin\omega t)^2 = \text{ave}\{R\cos(\omega t + \phi)\}^2$$

$$= R^2\text{ave}\cos(\omega t + \phi)^2$$

$$= \frac{R^2}{2}.$$

The quantity

$$\frac{R^2/2}{v} = \frac{\text{mean square value of signal}}{\text{mean square value of noise}},$$

called the *signal-to-noise ratio* or snr, indicates how well the signal shows up in the noise. The variance of $\tilde{\omega}$ may be rewritten as

$$\text{var } \tilde{\omega} \cong \frac{12}{n^3\,\text{snr}},$$

which shows, somewhat surprisingly, that a long series is more important than a strong signal.

For the variable-star data, these formulas yield standard deviations for the two frequencies of 1.04×10^{-5} (for frequency 0.2167) and 1.49×10^{-5} (for frequency 0.2618). These values show that the frequencies are in theory capable of very sharp resolution. It should be noted, however, that this result depends heavily on the assumptions made, especially independence of the errors.

The variances of $\tilde{A}$ and $\tilde{B}$ are increased by replacing ω by its estimate $\tilde{\omega}$. The results are

$$\operatorname{var} \tilde{A} \cong \frac{2v}{n} \frac{A^2 + 4B^2}{R^2},$$

$$\operatorname{var} \tilde{B} \cong \frac{2v}{n} \frac{4A^2 + B^2}{R^2},$$

$$\operatorname{cov}(\tilde{A}, \tilde{B}) \cong \frac{6v}{n} \frac{AB}{R^2},$$

$$\operatorname{cov}(\tilde{A}, \tilde{\omega}) \cong \frac{12v}{n^2} \frac{B}{R^2},$$

$$\operatorname{cov}(\tilde{B}, \tilde{\omega}) \cong \frac{-12v}{n^2} \frac{A}{R^2}.$$

Furthermore, estimates concerning different frequencies are, to this order of approximation, uncorrelated. Since, as Walker shows, $\tilde{A}$, $\tilde{B}$, and $\tilde{\omega}$ are all approximately normally distributed, these results allow us to find confidence intervals for the corresponding parameters.

In the light of the (approximate) standard deviations of the two estimated frequencies, it is instructive to recall that their final estimates differed by many times these quantities from the first values, computed directly from the data containing both components (see Table 2.1). Since the two frequencies are fairly close, they interact or interfere with each other, and this effect dominates the statistical error, unless it is removed, as in the simultaneous estimation procedure of Section 2.4. Pisarenko (1973) has shown that when a series contains two very similar frequencies the above formulas for variances and covariances may not be valid. Although Pisarenko's results are for frequencies closer than those in the present data, they suggest that we should treat the standard deviations given above with some caution.

Exercise 2.7 *The Estimates* $\bar{x}$, $\tilde{A}$, *and* $\tilde{B}$

Suppose that the data $\{x_0, \ldots, x_{n-1}\}$ were generated by the model (7). Show that

$$\bar{x} = \mu + \bar{a} + \left\{ A \cos \frac{(n-1)\omega}{2} + B \sin \frac{(n-1)\omega}{2} \right\} D_n(\omega),$$

where

$$D_n(\omega) = \frac{\sin n\omega/2}{n \sin \omega/2}.$$

Show also that

$$\tilde{A} = A + \frac{2}{n} \sum a_t \cos \omega t + \left\{ A \cos(n-1)\omega + B \sin(n-1)\omega \right\} D_n(2\omega)$$

$$- 2\bar{a} \cos \frac{(n-1)\omega}{2} D_n(\omega),$$

$$\tilde{B} = B + \frac{2}{n} \sum a_t \sin \omega t + \left\{ A \sin(n-1)\omega + B \cos(n-1)\omega \right\} D_n(2\omega)$$

$$- 2\bar{a} \sin \frac{(n-1)\omega}{2} D_n(\omega).$$

Exercise 2.8 *Continuation*

The bias of $\bar{x}$ as an estimator of μ is

$$b = \left\{ A \cos \frac{(n-1)\omega}{2} + B \sin \frac{(n-1)\omega}{2} \right\} D_n(\omega),$$

since $E\bar{a} = 0$. Show that

$$|b| \leqslant \frac{R}{n \sin \omega/2},$$

where, as usual, $R^2 = A^2 + B^2$. Find the bias of $\tilde{A}$ and show that it may similarly be bounded by

$$\frac{R}{n} \left\{ \frac{1}{\sin \omega} + \frac{2}{n(\sin \omega/2)^2} \right\} + \frac{2\mu}{n \sin \omega/2}.$$

Show that the same quantity is a bound for the bias of $\tilde{B}$.

Exercise 2.9   Continuation

Show that the variance of $\bar{x}$ is v/n, and that the variances of $\tilde{A}$ and $\tilde{B}$ differ from $2v/n$ by, at most, $(2v/n)/(n\sin\omega)$. Find the covariance of A and B, and show that it too is bounded by this same quantity.

Show that the covariances of $\bar{x}$ with $\tilde{A}$ and $\tilde{B}$ are both bounded by $(2v/n)/(n\sin\omega/2)$.

APPENDIX

The following program was used to fit the two-component model discussed in Section 2.4 and may also serve to fit the more general m-component model. Subprograms OPTOM, LOCALM, SSREG, STATS, and PARMS carry out the actual fitting algorithm. The main program and the subroutine DATIN are used solely for input/output.

These programs may also be used to fit the single-frequency model of Section 2.3, as a special case of the general model. The exact least squares method is found by setting APPFLG to .FALSE., and LIM and CONV to values that allow the algorithm to iterate to convergence (say, CONV $=1E-5$ and LIM$=5$). The approximate method is used by setting AP-PFLG to .TRUE., and LIM and CONV so that only one cycle is performed (that is, LIM$=1$ or CONV set to a relatively large value, say π).

NOTE: If any frequency is initialized at a value outside the range $[0,\pi]$, the final value will likewise fail to be its principal alias.

Subprogram LOCALM is from Richard P. Brent, *Algorithms for Minimization without Derivatives*, © 1973, pp. 188–190. Reprinted by permission of Prentice-Hall, Inc., Englewood Cliffs, New Jersey. Three statements have been changed and one has been added for the present application.

```
C
C      THIS MAIN PROGRAM AND SUBROUTINE DATIN ARE USED TO
C      INPUT DATA FOR, AND OUTPUT RESULTS FROM, FITTING A
C      MODEL OF HIDDEN PERIODICITIES TO A TIME SERIES.
C      THE TIME SERIES DATA COME FIRST IN THE INPUT FILE.
C      (FOR THE FORMAT SEE DATIN.) THE FIRST CARD AFTER THE
C      TIME SERIES CONTAINS THE VALUES OF THE VARIABLES
C      CONV, LIM AND APPFLG, WHICH CONTROL THE OPERATION
C      OF THE FITTING ALGORITHM, IN F10.5,I5,L1 FORMAT.
C      NEXT COMES A CARD WITH THE NUMBER OF COMPONENTS, IN
C      I5 FORMAT.
C      FOLLOWING THIS, THERE IS ONE CARD FOR EACH FREQUENCY
C      TO BE FITTED. IT CONTAINS THE STARTING VALUES OF THE
C      FREQUENCY AND THE COSINE AND SINE COEFFICIENTS, IN
C      3F10.5 FORMAT.
C
       DIMENSION X(600),FRE(10),A(10),B(10)
       LOGICAL APPFLG
       CALL DATIN (X,N,START,STEP,5)
       READ (5,1) CONV,LIM,APPFLG
1      FORMAT (F10.5,I5,L1)
       WRITE(6,4) CONV,LIM,APPFLG
4      FORMAT (≠OFOR THIS RUN,  CONV =≠,E12.4/
      +        ≠                 LIM =≠,I5/
      +        ≠               APPFLG =≠,L5)
       READ(5,8) M
8      FORMAT(I5)
       DO 10 J=1,M
10     READ (5,2) FRE(J),A(J),B(J)
2      FORMAT (3F10.5)
       WRITE(6,5)
5      FORMAT(≠OINITIAL VALUES ARE -≠)
       WRITE(6,3) (J,FRE(J),A(J),B(J),J=1,M)
3      FORMAT(≠ COMPONENT FREQUENCY      COSINE       SINE≠/
      +       ≠                         COEFFICIENTS≠/
      +       (I5,F15.7,2E15.6))
       CALL OPTOM (X,N,RMU,FRE,A,B,M,CONV,LIM,APPFLG)
       WRITE(6,6)
6      FORMAT(≠O  FINAL VALUES ARE -≠)
       WRITE(6,3) (IM,FRE(IM),A(IM),B(IM),IM=1,M)
       SS=0.0
       DO 30 I=1,N
       TEMP=X(I)-RMU
       DO 40 J=1,M
       ARG=FLOAT(I-1)*FRE(J)
       TEMP=TEMP-A(J)*COS(ARG)-B(J)*SIN(ARG)
40     CONTINUE
       SS=SS+TEMP**2
30     CONTINUE
       WRITE(6,7) RMU,SS
7      FORMAT(≠OFITTED CONSTANT IS          ≠,E15.6/
      +        ≠ RESIDUAL SUM OF SQUARES IS≠,E15.6)
       STOP
       END
```

```
      SUBROUTINE DATIN (X,N,START,STEP,M)
C
C   THIS SUBROUTINE IS USED TO INPUT A SERIES OF VALUES
C   (IN RUN-TIME FORMAT) AND SOME ASSOCIATED QUANTITIES
C   (IN FIXED FORMAT).    THE FIRST FOUR DATA CARDS ARE -
C   1   A HEADER CARD   (72 COLUMNS)
C   2   VALUE OF N   (I5)
C   3   THE DATA FORMAT (72 COLUMNS)
C   4   START AND STEP   (2F10.5)
C   PARAMETERS ARE -
C
C NAME    TYPE                          VALUE
C                           ON ENTRY            ON RETURN
C
C X     REAL ARRAY NOT USED            THE SERIES
C
C N      INTEGER    NOT USED           SERIES LENGTH
C
C START REAL       NOT USED            TIME VALUE AT THE
C                                      FIRST DATA POINT
C
C STEP  REAL       NOT USED            TIME INCREMENT
C                                      BETWEEN DATA POINTS
C
C M      INTEGER    LOGICAL UNIT NUMBER   UNCHANGED
C
      DIMENSION X(600),HEAD(18),FMT(18)
      READ(M,1) HEAD,N,FMT,START,STEP
    1 FORMAT(18A4/I5/18A4/2F10.5)
      WRITE(6,2) HEAD,N,FMT,START,STEP
    2 FORMAT(≠0THE DATA HEADER READS -≠/1X,18A4/
     +        ≠ THE SERIES LENGTH IS≠,I6/
     +        ≠ THE DATA FORMAT IS -≠/1X,18A4/
     +        ≠ TIME ORIGIN IS≠,F11.5,
     +        ≠,  TIME INCREMENT IS≠,F11.5)
      READ(M,FMT) (X(I),I=1,N)
      RETURN
      END
```

```
              SUBROUTINE OPTOM (X,N,RMU,FRE,A,B,M,CONV,LIM,APPFLG)
C
C     THIS SUBROUTINE,   WITH SUBPROGRAMS LOCALM,   SSREG,
C     STATS AND PARMS,   IMPLEMENTS THE ALGORITHM
C     FOR THE LEAST-SQUARES FITTING OF THE
C     MODEL OF HIDDEN PERIODICITIES.   PARAMETERS ARE
C
C     NAME    TYPE                              VALUE
C                               ON ENTRY                ON RETURN
C
C     X       REAL ARRAY    THE TIME SERIES            UNCHANGED
C
C     N       INTEGER       SERIES LENGTH              UNCHANGED
C
C     RMU     REAL          NOT USED                   CONSTANT TERM
C
C     FRE     REAL ARRAY    STARTING VALUES FOR        FINAL VALUES
C                           THE FREQUENCIES TO
C                           BE FITTED
C
C     A       REAL ARRAY    STARTING VALUES FOR        FINAL VALUES
C                           COSINE COEFFICIENTS
C
C     B       REAL ARRAY    STARTING VALUES FOR        FINAL VALUES
C                           SINE COEFFICIENTS
C
C     M       INTEGER       NUMBER OF COMPONENTS       UNCHANGED
C                           TO BE FITTED
C
C     CONV REAL             CONVERGENCE CRITERION      UNCHANGED
C                           ITERATION CEASES WHEN
C                           IN ONE CYCLE,   NO
C                           FREQUENCY CHANGES BY
C                           MORE THAN CONV
C
C     LIM     INTEGER       MAXIMUM NUMBER OF          UNCHANGED
C                           CYCLES OF ITERATION
C
C     APPFLG LOGICAL        FLAG CONTROLLING           UNCHANGED
C                           WHETHER APPROXIMATE
C                           (.TRUE.)   OR EXACT
C                           (.FALSE.)   LEAST SQUARES
C                           IS TO BE USED
C
```

```
C     NOTES.
C     1.   IF VALUES OF N OR M EXCEEDING 600 AND 10,
C          RESPECTIVELY,  ARE USED,  THE DIMENSION
C          STATEMENT BELOW SHOULD BE CHANGED ACCORDINGLY.
C     2.   USE OF EXACT LEAST SQUARES  (APPFLG = .FALSE.)
C          WILL CAUSE LONGER EXECUTION TIMES.
C     3.   THE FREQUENCIES IN GENERAL CONVERGE TO VALUES
C          WITHIN 2*PI/N OF THEIR STARTING VALUES.  THE
C          STARTING VALUES SHOULD BE GIVEN TO AT LEAST THIS
C          ACCURACY,  OR THE ALGORITHM MAY BE TRAPPED BY
C          SIDE-LOBES.  STARTING VALUES OF THE COSINE AND
C          SINE COEFFICIENTS ARE LESS CRITICAL,  AND MAY BE
C          SET TO ZERO.
C
      REAL LOCALM
      DIMENSION X(N),Y(600),FRE(10),A(10),B(10)
      LOGICAL APPFLG
      DATA EPS /1E-9/
      T=CONV
      DELTA =3.142/FLOAT(N)
      DO 10 KOUNT=1,LIM
      SUM=0
      DO 20 I=1,N
      Y(I)=X(I)
      DO 30 J=1,M
      ARG=FLOAT(I-1)*FRE(J)
30    Y(I)=Y(I)-A(J)*COS(ARG)-B(J)*SIN(ARG)
20    SUM=SUM+Y(I)
      RMU=SUM/FLOAT(N)
      TEST=0
      DO 40 J=1,M
      DO 50 I=1,N
      Y(I)=X(I)-RMU
      DO 50 K=1,M
      IF(K .EQ.J) GO TO 50
      ARG=FLOAT(I-1)*FRE(K)
      Y(I)=Y(I)-A(K)*COS(ARG)-B(K)*SIN(ARG)
50    CONTINUE
      DUMMY=LOCALM (FRE(J)-DELTA,FRE(J)+DELTA,
     +     EPS,T,TEMP,Y,N,APPFLG)
      TEST=AMAX1(TEST,ABS(FRE(J)-TEMP))
      FRE(J)=TEMP
40    CALL PARMS (Y,N,FRE(J),APPFLG,A(J),B(J))
      IF (TEST .LT. CONV) RETURN
      DELTA=TEST+2.0*T
10    CONTINUE
      RETURN
      END
```

```
          REAL FUNCTION LOCALM (A,B,EPS,T  ,X,Y,N,APPFLG)
C         REAL FUNCTION LOCALM (A,B,EPS,T,F,X)
C
C     THIS IS THE FORTRAN FUNCTION LOCALM GIVEN BY
C     RICHARD BRENT IN
C           ALGORITHMS FOR MINIMIZATION WITHOUT DERIVATIVES
C     (PRENTICE-HALL, 1973).
C     IT FINDS A LOCAL MINIMUM OF THE FUNCTION F IN THE
C     INTERVAL (A,B).
C     T AND EPS DEFINE A TOLERANCE  TOL = EPS*ABS(X)+T,
C     WHERE  X  IS THE CURRENT APPROXIMATION TO THE POSITION
C     OF THE MINIMUM.  THE MINIMUM IS FOUND WITH AN ERROR
C     OF AT MOST  3*TOL.
C     F IS NOT EVALUATED AT POINTS CLOSER THAN TOL.
C     A SUITABLE VALUE FOR  EPS  IS THE SQUARE ROOT OF THE
C     RELATIVE MACHINE PRECISION.  FOR MORE DETAILS SEE THE
C     ABOVE REFERENCE.
C
      DIMENSION Y(N)
      REAL M
      SA=A
      SB=B
      X=SA+0.381966*(SB-SA)
      W=X
      V=W
      E=0.0
C     FX=F(X)
      FX=-SSREG(Y,N,X,APPFLG)
      FW=FX
      FV=FW
  10  M=0.5*(SA+SB)
      TOL=EPS*ABS(X)+T
      T2=2.0*TOL
      IF(ABS(X-M) .LE. T2-0.5*(SB-SA)) GO TO 190
      R=0.0
      Q=R
      P=Q
      IF (ABS(E) .LE. TOL) GO TO 40
      R=(X-W)*(FX-FV)
      Q=(X-V)*(FX-FW)
      P=(X-V)*Q-(X-W)*R
      Q=2.0*(Q-R)
      IF (Q .LE. 0.0) GO TO 20
      P=-P
      GO TO 30
  20  Q=-Q
  30  R=E
      E=D
  40  IF (ABS(P) .GE. ABS(0.5*Q*R)) GO TO 60
      IF ((P .LE. Q*(SA-X)) .OR. (P .GE. Q*(SB-X))) GO TO 60
      D=P/Q
      U=X+D
      IF((U-SA .GE. T2) .AND. (SB-U .GE. T2)) GO TO 90
      IF (X .GE. M) GO TO 50
      D=TOL
```

37

```
            GO TO 90
    50      D=-TOL
            GO TO 90
    60      IF (X .GE. M) GO TO 70
            E=SB-X
            GO TO 80
    70      E=SA-X
    80      D=0.381966*E
    90      IF (ABS(D) .LT. TOL) GO TO 100
            U=X+D
            GO TO 120
   100      IF (D .LE. 0.0) GO TO 110
            U=X+TOL
            GO TO 120
   110      U=X-TOL
  C120      FU=F(U)
   120      FU=-SSREG(Y,N,U,APPFLG)
            IF (FU .GT. FX) GO TO 150
            IF (U .GE. X) GO TO 130
            SB=X
            GO TO 140
   130      SA=X
   140      V=W
            FV=FW
            W=X
            FW=FX
            X=U
            FX=FU
            GO TO 10
   150      IF (U .GE. X) GO TO 160
            SA=U
            GO TO 170
   160      SB=U
   170      IF ((FU .GT. FW) .AND. (W .NE. X)) GO TO 180
            V=W
            FV=FW
            W=U
            FW=FU
            GO TO 10
   180      IF ((FU .GT. FV) .AND. (V .NE. X) .AND. (V .NE. W))
           +      GO TO 10
            V=U
            FV=FU
            GO TO 10
   190      LOCALM=FX
            RETURN
            END
```

```
      FUNCTION SSREG (Y,N,OMEGA,APPFLG)
C
C   THIS FUNCTION RETURNS THE SUM  OF SQUARES  (APPROXIMATE
C   OR EXACT)  ASSOCIATED WITH OMEGA.  PARAMETERS ARE
C
C   NAME    TYPE        VALUE ON ENTRY  (NONE ARE CHANGED)
C
C   Y    REAL ARRAY  THE TIME SERIES
C
C   N      INTEGER    SERIES LENGTH
C
C   OMEGA REAL        THE FREQUENCY
C
C   APPFLG LOGICAL    .TRUE.  FOR APPROXIMATE LEAST SQUARES,
C                     .FALSE.FOR EXACT LEAST SQUARES
C
      DIMENSION Y(N)
      LOGICAL APPFLG
      CALL STATS (Y,N,OMEGA,CY,SY)
      IF (APPFLG) GO TO 10
      RN=N
      CON=SIN(RN*OMEGA)/SIN(OMEGA)
      ARG=(RN-1.0)*OMEGA
      CC=0.5*(RN+COS(ARG)*CON)
      CS=0.5*SIN(ARG)*CON
      SS=RN-CC
      SSREG=(SS*CY**2-2.0*CS*CY*SY+CC*SY**2)/(CC*SS-CS**2)
      RETURN
   10 CONTINUE
      SSREG=(CY**2+SY**2)*2.0/FLOAT(N)
      RETURN
      END
```

```
      SUBROUTINE STATS (Y,N,OMEGA,CY,SY)
C
C    THIS SUBROUTINE RETURNS THE COSINE AND SINE SUMS OF A
C    TIME SERIES.   PARAMETERS ARE
C
C    NAME     TYPE                          VALUE
C                              ON ENTRY            ON RETURN
C
C    Y       REAL ARRAY THE TIME SERIES     UNCHANGED
C
C    N       INTEGER    SERIES LENGTH       UNCHANGED
C
C    OMEGA REAL         THE FREQUENCY       UNCHANGED
C
C    CY      REAL       NOT USED            COSINE SUM
C
C    SY      REAL       NOT USED            SINE SUM
C
      DIMENSION Y(N)
      CY=0.0
      SY=0.0
      DO 10 I=1,N
      ARG=FLOAT(I-1)*OMEGA
      CY=CY+COS(ARG)*Y(I)
      SY=SY+SIN(ARG)*Y(I)
 10   CONTINUE
      RETURN
      END
```

```
      SUBROUTINE PARMS (Y,N,OMEGA,APPFLG,A,B)
C
C   THIS SUBROUTINE RETURNS THE  (EXACT OR APPROXIMATE)
C   LEAST SQUARES ESTIMATES OF THE COSINE AND SINE
C   COEFFICIENTS OF A SINGLE PERIODIC COMPONENT.
C   PARAMETERS ARE
C
C  NAME      TYPE                          VALUE
C                             ON ENTRY              ON RETURN
C
C  Y      REAL ARRAY THE TIME SERIES       UNCHANGED
C
C  N       INTEGER    SERIES LENGTH         UNCHANGED
C
C  OMEGA  REAL       THE FREQUENCY         UNCHANGED
C
C  APPFLG LOGICAL    .TRUE.  OR  .FALSE.   UNCHANGED
C                    FOR APPROXIMATE OR
C                    EXACT LEAST SQUARES,
C                    RESPECTIVELY
C
C  A      REAL       NOT USED             COS COEFFICIENT
C
C  B      REAL       NOT USED             SIN COEFFICIENT
C
      LOGICAL APPFLG
      DIMENSION Y(N)
      CALL STATS (Y,N,OMEGA,CY,SY)
      RN=FLOAT(N)
      IF (SIN(OMEGA) .EQ. 0.0) GO TO 20
      IF (APPFLG) GO TO 10
      CON=SIN(RN*OMEGA)/SIN(OMEGA)
      ARG=(RN-1.0)*OMEGA
      CC=0.5*(RN+COS(ARG)*CON)
      CS=0.5*SIN(ARG)*CON
      SS=RN-CC
      DEL=CC*SS-CS**2
      A=(CY*SS-SY*CS)/DEL
      B=(SY*CC-CY*CS)/DEL
      RETURN
   10 CONTINUE
      A=2.0*CY/RN
      B=2.0*SY/RN
      RETURN
   20 A=CY/RN
      B=0.0
      RETURN
      END
```

3

HARMONIC ANALYSIS

In Chapter 2 we discussed how to find the parameters A, B, and ω of a cosine function $A\cos\omega t + B\sin\omega t$, using least squares. This procedure is useful when periodicity clearly exists in the data and needs to be described exactly. In this chapter, we shall describe a method that may be used to analyze an arbitrary set of data into periodic components whether or not the data appear periodic.

There are, of course, sets of data for which such an analysis is meaningless. However, other sets of data, although they may not appear to be periodic, do in fact contain interesting periodic components. It is in detecting such components that harmonic analysis is irreplaceable.

3.1 THE FOURIER FREQUENCIES

The central result that underlies harmonic analysis is the following set of identities. Let $\omega_j = 2\pi j / n$, the jth *Fourier frequency*. Because of aliasing (see Section 2.5), we need consider only frequencies satisfying $0 \le \omega_j \le \pi$, which corresponds to $0 \le j \le n/2$. Note that, if n is even, $\omega_{n/2} = \pi$, but if n is odd, π is not a Fourier frequency. The identities follow from the results of Exercise 2.2 and are as follows:

$$\sum_{t=0}^{n-1} \cos \omega_j t = 0, \qquad j \ne 0,$$

$$\sum_{t=0}^{n-1} \sin \omega_j t = 0,$$

$$\sum_{t=0}^{n-1} \cos \omega_j t \cos \omega_k t = \begin{cases} n/2, & j=k\neq 0 \text{ or } n/2, \\ n, & j=k=0 \text{ or } n/2, \\ 0, & j\neq k, \end{cases} \qquad (1)$$

$$\sum_{t=0}^{n-1} \cos \omega_j t \sin \omega_k t = 0,$$

$$\sum_{t=0}^{n-1} \sin \omega_j t \sin \omega_k t = \begin{cases} n/2, & j=k\neq 0 \text{ or } n/2, \\ 0, & \text{otherwise.} \end{cases}$$

The special results for $j=k=0$ or $n/2$ hold essentially because the sine terms vanish identically at frequencies 0 and π. The first two relations are, in fact, special cases of the next two, with j set equal to 0, and k replaced by j. The results imply that the sines and cosines of the Fourier frequencies are *orthogonal* (with respect to summation over the integers $0,\ldots,n-1$). (It is an important fact that they are also orthogonal with respect to integration. See Exercise 3.1.)

Now suppose that $x_0,\ldots,x_{n-1}$ are any set of n numbers. Let

$$A_0 = \frac{1}{n}\sum_{t=0}^{n-1} x_t = \bar{x},$$

$$A_j = \frac{2}{n}\sum_{t=0}^{n-1} x_t \cos \omega_j t, \qquad (2)$$

$$B_j = \frac{2}{n}\sum_{t=0}^{n-1} x_t \sin \omega_j t,$$

for $0<j<n/2$, and if n is even,

$$A_{n/2} = \frac{1}{n}\sum_{t=0}^{n-1} x_t \cos \omega_{n/2} t$$

$$= \frac{1}{n}\sum_{t=0}^{n-1} (-1)^t x_t.$$

Then it may be seen from the orthogonality relations (1) that

$$x_t = A_0 + \sum_{0 < j < n/2} \left(A_j \cos \omega_j t + B_j \sin \omega_j t \right)$$

$$+ (-1)^t A_{n/2}, \qquad t = 0, \ldots, n-1, \tag{3}$$

the last term being included only if n is even (see Exercise 3.3). Thus we can represent an arbitrary sequence of numbers as a sum of periodic components. Notice that, for n even or odd, there are n coefficients in the sum.

This, of course, is not the only way in which such a representation can be found. If we regard the n data values $x_0, \ldots, x_{n-1}$ as a single point in n-dimensional Euclidean space, a similar representation will hold for any set of frequencies whose cosines and sines form a basis. However, the Fourier frequencies are a natural set to use, being equally spaced over the range of frequencies we wish to employ. Also, because of the orthogonality relations (1), calculation of the coefficients $\{A_0, A_1, B_1, \ldots, A_{n/2}\}$ is straightforward. This would not be true for most other sets of frequencies.

Equation 3 has been shown to hold only for t in the range $0, \ldots, n-1$. Any other value of t may be written as $kn + t'$, for some nonzero integer k and t' within the required range. If we substitute this value $kn + t'$ for t in the right-hand side of (3), the k's drop out because of the form of the Fourier frequencies. Thus the value found is simply $x_{t'}$. Hence, if t is an unrestricted integer variable, the sum defines a periodic sequence of period n, consisting of the values $x_0, \ldots, x_{n-1}$ repeated cyclically.

The jth Fourier frequency has period $2\pi / \omega_j = n/j$. A sinusoid with this frequency executes j complete cycles in the span of the data, thus providing a useful interpretation of the index j. One consequence of this is that few of the periods are integers (although of course all are rational). The harmonic analysis (3) represents the decomposition of a series into components each of which is repeated a whole number of times in the span of the data.

Exercise 3.1 Orthogonality of Sinusoids with Respect to Integration

(i) If n is an integer, show that

$$\int_{-\pi}^{\pi} \cos nx \, dx = \begin{cases} 2\pi & \text{if } n = 0, \\ 0 & \text{otherwise,} \end{cases}$$

and

$$\int_{-\pi}^{\pi} \sin nx \, dx = 0, \qquad \text{all } n.$$

(ii)   Hence show that, for integers m, n,

$$\int_{-\pi}^{\pi} \cos mx \cos nx\, dx = \begin{cases} 2\pi, & m = n = 0, \\ \pi, & m = n \neq 0, \\ 0, & m \neq n, \end{cases}$$

$$\int_{-\pi}^{\pi} \cos mx \sin nx\, dx = 0, \qquad \text{all } m, n,$$

$$\int_{-\pi}^{\pi} \sin mx \sin nx\, dx = \begin{cases} \pi, & m = n \neq 0, \\ 0, & \text{otherwise.} \end{cases}$$

Exercise 3.2   Orthogonality with Respect to Midpoints

Suppose that the interval $(-\pi, \pi)$ is divided into N equal subintervals. Denote the midpoints of these intervals by x_j, $j = 1, \ldots, N$.

(i)   For integer n, show that

$$\frac{1}{N} \sum_{j=1}^{N} \cos nx_j = \begin{cases} 1 & \text{if } n \equiv 0 \ (modulo\ N), \\ 0 & \text{otherwise,} \end{cases}$$

and

$$\frac{1}{N} \sum_{j=1}^{N} \sin nx_j = 0, \qquad \text{all } n.$$

(ii)   Deduce the orthogonality relations corresponding to those found in Exercise 3.1.

Exercise 3.3   Transposed Orthogonality Relations

Since $\omega_j t = 2\pi jt/n = \omega_t j$, the orthogonality relations (1) may also be written (for t and u in the range 0 to $n-1$) as

$$\cos \omega_0 t \cos \omega_0 u + 2 \sum_{0 < j < n/2} \cos \omega_j t \cos \omega_j u + \cos \omega_{n/2} t \cos \omega_{n/2} u$$

$$= \begin{cases} n/2, & t = u \neq 0, \\ n, & t = u = 0, \\ 0, & t \neq u, \end{cases}$$

$$2 \sum_{0 < j < n/2} \cos \omega_j t \sin \omega_j u = 0, \qquad \text{all } t, u,$$

$$2 \sum_{0 < j < n/2} \sin \omega_j t \sin \omega_j u = \begin{cases} n/2, & t = u \neq 0, \\ 0, & \text{otherwise;} \end{cases}$$

here any term with the subscript $n/2$ is omitted if n is odd. Use these relations to verify the inversion formula (3).

3.2   COMPLEX-VALUED DATA: THE DISCRETE FOURIER TRANSFORM

The theory of Section 3.1 is simpler in terms of complex-valued data. There are, in fact, occasions when a pair of real-valued time series can most naturally be regarded as a single complex-valued series. An example analyzed by Brillinger (1973) is the deviation of the axis of instantaneous rotation of the earth. The deviation is measured by a pair of coordinates, which may be treated theoretically as a single complex number.

In general we are faced with strictly real-valued data. These can always be regarded as complex numbers with zero imaginary parts, although this may seem an unnecessary complication. However, certain algebraic simplifications that arise make the required stretch of the imagination seductively appealing.

We have already used the Euler relation

$$\cos\lambda + i\sin\lambda = \exp(i\lambda).$$

This, together with the related formulas

$$\cos\lambda = \frac{\exp(i\lambda) + \exp(-i\lambda)}{2},$$

$$\sin\lambda = \frac{\exp(i\lambda) - \exp(-i\lambda)}{2i},$$

shows that the complex exponential function $\exp(i\lambda)$ is closely related to the cosines and sines. The orthogonality relations corresponding to (1.1) are[†]

$$\sum_{t=0}^{n-1} \exp(i\omega_j t)\exp(i\omega_k t)^* = \sum_{t=0}^{n-1} \exp(i\omega_j t)\exp(-i\omega_k t)$$

$$= \begin{cases} n & \text{if } j \equiv k \ (mod \ n), \\ 0 & \text{otherwise.} \end{cases} \tag{4}$$

The result for $j \equiv k \ (mod\,n)$ is immediate, for then each term is 1. The other values follow from Exercise 2.2.

[†]The asterisk is used to denote complex conjugation.

Now suppose that $x_0, \ldots, x_{n-1}$ are any set of n complex numbers. Let

$$J_j = \frac{1}{n} \sum_{t=0}^{n-1} x_t \exp(-i\omega_j t). \tag{5}$$

Notice that $J_j = \frac{1}{2}(A_j - iB_j)$. The data may be recovered from the equation

$$x_t = \sum J_j \exp(i\omega_j t), \tag{6}$$

which follows from the transposed version of (4) (see Exercise 3.4). The summation here is for j in either of the ranges $-n/2 < j \leqslant n/2$ and $0 \leqslant j < n$. Both ranges give the same answer because (2) defines a sequence of values that are periodic in j with period n. The range $-n/2 < j \leqslant n/2$ is the most natural one in light of the preceding section. The range of frequencies remains the same as in that section, except that negative frequencies have to be added, since $\exp(i\omega t)$ and $\exp(-i\omega t)$ are not identical. However, if we write the range of summation as $j = 0, \ldots, n-1$, (5) and (6) become very similar.

By analogy with the usual (integral) Fourier transform, $J_0, \ldots, J_{n-1}$ is called the *discrete Fourier transform* of $x_0, \ldots, x_{n-1}$, which are obtained from $J_0, \ldots, J_{n-1}$ by the *inverse transform* (6). In the same way, we shall refer to the A's and B's of Section 3.1 as the cosine and sine transforms, respectively.

We do not gain anything other than this algebraic simplicity by regarding real data as complex. From (5),

$$J_{n-j} = \frac{1}{n} \sum_{t=0}^{n-1} x_t \exp(-i\omega_{n-j} t)$$

$$= \frac{1}{n} \sum_{t=0}^{n-1} x_t \exp(i\omega_j t),$$

since $\omega_{n-j} t = \{2\pi(n-j)/n\} t = 2\pi t - (2\pi j/n)t = 2\pi t - \omega_j t$. But if all the x's are real, then

$$J_{n-j} = \frac{1}{n} \sum_{t=0}^{n-1} x_t \exp(i\omega_j t) = J_j^*.$$

These constraints mean that only half the J's are free to vary independently, and in fact they are all determined by the n A's and B's defined in Section 3.1.

Since the discrete Fourier transform is a set of complex numbers, there are two natural ways to represent it. The first is in terms of real and imaginary parts, which are just one-half the cosine and sine coefficients A_j and B_j of the preceding section. The second is in terms of its *magnitude* and *phase*, R_j and ϕ_j, where

$$J_j = R_j \exp(i\phi_j).$$

The cosine and sine coefficients A_j and B_j are the same as the simple estimates of similar quantities described in Section 2.2. In fact, since we are now dealing only with Fourier frequencies, they are also the same as the other estimates, $\hat{A}$ and $\hat{B}$, and $\tilde{A}$ and $\tilde{B}$, respectively. There is a minor notational conflict, in that

$$R_j^2 = |J_j|^2 = J_j J_j^*$$

$$= \tfrac{1}{4}(A_j + iB_j)(A_j - iB_j)$$

$$= \tfrac{1}{4}(A_j^2 + B_j^2) = \tfrac{1}{4}\tilde{R}^2.$$

The same function will arise later as the *periodogram*, which is defined as

$$I(\omega_j) = \frac{n}{2\pi} R_j^2.$$

As in Section 2.3, we shall also refer to R_j^2 as the periodogram when the different scaling does not matter.

An important property of the Fourier transform, and one that will be used in many subsequent arguments, is known as *linearity* or *superposability*. It is an elementary property, but to emphasize the point we demonstrate it here. Suppose that the data $\{x_t\}$ are the sums (or *superposition*) of $\{y_t\}$ and $\{z_t\}$. Then

$$\sum_t x_t \exp(i\omega_j t) = \sum_t (y_t + z_t)\exp(i\omega_j t)$$

$$= \sum_t y_t \exp(i\omega_j t) + \sum_t z_t \exp(i\omega_j t),$$

or, in an obvious notation,

$$J_{x,j} = J_{y,j} + J_{z,j}.$$

In words, the transform of the sum is the sum of the transforms.

In the rest of this book, we shall use the complex version of the discrete Fourier transform extensively. However, unless stated otherwise, the series being analyzed will *always* be assumed to be real.

Exercise 3.4   *Transposed Orthogonality Relations*

The transposed version of (4) is

$$\sum_j \exp(i\omega_j t)\exp(-i\omega_j u) = \begin{cases} n & \text{if } t \equiv u \,(mod\ n), \\ 0 & \text{otherwise}, \end{cases}$$

where the range of summation is any set of n consecutive values of j. Verify this, and use it to derive the inverse transform (6).

Exercise 3.5   *A Generalized Transform*

It is often convenient to define the discrete Fourier transform for *any* frequency ω as

$$J(\omega) = \frac{1}{n} \sum_{t=0}^{n-1} x_t \exp(-i\omega t).$$

In this notation, $J_j = J(\omega_j)$. There are two further inversion formulas involving this function.

(i) Derive the complex analogs of the integral orthogonality relations derived in Exercise 3.1. Deduce that

$$\frac{n}{2\pi} \int_{-\pi}^{\pi} J(\omega)\exp(it\omega)\,d\omega = \begin{cases} x_t, & 0 \leqslant t < n, \\ 0, & \text{otherwise}. \end{cases}$$

(ii) The midpoint orthogonality relations derived in Exercise 3.2 have a more general complex analog

$$\sum_{j=0}^{n-1} \exp\{it(\omega_j+\theta)\}\exp\{-iu(\omega_j+\theta)\} = \begin{cases} n, & t \equiv u\,(mod\ n), \\ 0 & \text{otherwise}. \end{cases}$$

Verify this, and deduce that

$$\sum_{j=0}^{n-1} J(\omega_j+\theta)\exp\{i(\omega_j+\theta)\} = x_t, \qquad 0 \leqslant t < n.$$

Note that it is sufficient to consider θ in the range $0 < \theta < \omega_1$. What values are found if the sum is evaluated for other values of t?

Parts (i) and (ii) both provide inversion formulas for obtaining $x_0, \ldots, x_{n-1}$ from the transform $J(\omega)$, $-\pi < \omega \leqslant \pi$ or $0 \leqslant \omega < 2\pi$. Note, however, that these formulas give different answers when evaluated for t outside this range.

3.3   DECOMPOSING THE SUM OF SQUARES

The orthogonality relations of Sections 3.1 and 3.2 imply identities between the sums of squares of the original data and of the transforms. For real data, this takes the form

$$\sum_{t=0}^{n-1} x(t)^2 = nA_0^2 + \frac{n}{2} \sum_{0<j<n/2} \left(A_j^2 + B_j^2\right) + nA_{n/2}^2. \tag{7}$$

The final term is included only if n is even. Equation 7 may also be written as

$$\sum_{t=0}^{n-1} x(t)^2 = n\bar{x}^2 + 4\pi \sum_{0<j<n/2} I(\omega_j) + 8\pi I(\pi),$$

with the same proviso. For complex data, the corresponding form is

$$\sum_{t=0}^{n-1} |x_t|^2 = n \sum_{j=0}^{n-1} |J_j|^2, \tag{8}$$

which again shows the algebraic simplicity of the complex version.

This partitioning of a sum of squares is usually known as an analysis of variance. In the present case, it may be interpreted as parallel to representation (3) of the data. In (3), we exhibit the data as a sum of various periodic terms. In (7), we decompose the sum of squares of the data into components associated with the corresponding frequencies.

The same orthogonality relations mean that the least squares methods of Chapter 2 are especially simple if we may restrict our attention to the Fourier frequencies. In the first place, the Fourier frequencies are the frequencies for which the least squares estimates $\hat{A}$ and $\hat{B}$ are the same as the simpler estimates $\tilde{A}$ and $\tilde{B}$. More importantly, a standard feature of least squares is that, when fitting orthogonal variables, the results of fitting variables separately are the same as those for fitting simultaneously. This removes the biasing effect seen in Section 2.4, which arose because one of the frequencies in the data was not a Fourier frequency.

Exercise 3.6   Decomposing the Sum of Squares

Relation 1 may be derived either from (2) and the transposed orthogonality relations (Exercise 3.3), or from (3) and the original orthogonality relations (1). Verify this, and show that the complex version may similarly be derived in two ways.

3.4 HARMONIC ANALYSIS OF SOME SPECIAL FUNCTIONS

To display some of the properties of the discrete Fourier transform, we now examine the harmonic analysis of some simple functions.

(i) A Cosine Wave

In terms of Fourier analysis, the simplest place to begin is with a cosine wave, $R\cos(\omega t + \phi)$. Since the transform is a linear operation, R will occur only as a scale factor throughout, and so we may give it the value 1. Now

$$\cos(\omega t + \phi) = \tfrac{1}{2}\left[\exp\{i(\omega t + \phi)\} + \exp\{-i(\omega t + \phi)\}\right];$$

hence it is sufficient to find the transform of $\exp(i\omega t)$ [once again, $\exp(i\phi)$ is just a scale factor]. But the (complex) transform of $\exp(i\omega t)$ is

$$\frac{1}{n}\sum_{t=0}^{n-1}\exp(i\omega t)\exp(-i\omega_j t),$$

where $\omega_j = 2\pi j / n$ is the jth Fourier frequency. Using our previous results, we obtain for this sum

$$\frac{1}{n}\sum_{t=0}^{n-1}\exp\{i(\omega - \omega_j)t\} = \exp\left\{\frac{i(n-1)(\omega - \omega_j)}{2}\right\}\frac{\sin\{n(\omega - \omega_j)/2\}}{n\sin\{(\omega - \omega_j)/2\}}.$$

As we would expect, if $\omega = \omega_k$ is itself a Fourier frequency, the transform vanishes unless $j = k$, when it takes the value 1. If ω is not a Fourier frequency, all the terms are nonzero. The amplitudes of the terms are given by the modulus of the second factor, and the phases by the first.

The appearance of nonzero terms in the transform because of a sinusoid of a different frequency is called *leakage*. Since we carry out a harmonic analysis to *separate* the effects of different frequencies, leakage is a confusing phenomenon. Furthermore, the frequencies that are present in the series being analyzed are often not Fourier frequencies (although the series length n may sometimes be chosen so that they are), and hence leakage is the rule rather than the exception. Techniques for reducing leakage are discussed in Section 5.2.

The function $(\sin\omega n/2)/(n\sin\omega/2)$ has arisen before and will arise again. It is a version of the Dirichlet function (Titchmarsh, 1939, p. 402) and is denoted as $D_n(\omega)$. If n is moderately large, and ω small, it is approximately $(\sin\omega n/2)/(\omega n/2)$. A graph of the function $(\sin x)/x$ appears in Figure 3.1. Its principal features are the peak of height 1 at the origin, its zeros at nonzero multiples of π, and its relatively slowly decaying minor peaks, or *sidelobes*.

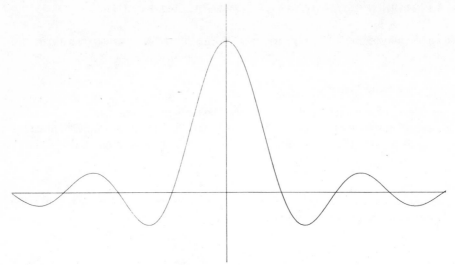

Figure 3.1   The function $(\sin x)/x$.

Thus the amplitude of the transform consists approximately of this function, centered around ω and rescaled so that the zeros are separated by $2\pi/n$, the distance between adjacent Fourier frequencies.

Hence, when ω is not a Fourier frequency, only the values of the transform that are close to ω will be large, and the values decay inversely proportionally to $|\omega - \omega_j|$.

The transform of the original cosine wave will consist of contributions from the frequencies ω and $-\omega$. If we concentrate on the positive frequencies, as we may for real data, the terms centered around $-\omega$ will have little effect unless ω is close to 0 or π.

(ii)   A Single Impulse

If we transform the sequence that consists of a single nonzero value, say at time t, of value 1, the transform is

$$J_j = \frac{1}{n} \exp(-it\omega_j).$$

It has constant amplitude, and the phase is (*mod* 2π) a linear function of frequency.

(iii)   A Step Function

Let

$$x_t = \begin{cases} 1 & \text{if } 0 \leqslant t < m, \\ 0 & \text{if } m \leqslant t < n, \end{cases}$$

for some m in the range $1, \ldots, n-2$. Then

$$J_j = \frac{1}{n} \sum_{t=0}^{n-1} x_t \exp(-i\omega_j t)$$

$$= \frac{1}{n} \sum_{t=0}^{m-1} \exp(-i\omega_j t),$$

and thus

$$J_j = \exp\left\{ \frac{-i\omega_j(m-1)}{2} \right\} \frac{\sin \omega_j m/2}{n \sin \omega_j/2}.$$

The amplitude of the transform is thus the Dirichlet function, (m/n) $D_m(\omega_j)$. Hence it decays roughly as $1/\omega_j$, from a maximum of m/n (which is less than 1) at zero frequency. As in examples (i) and (ii), the phase of the transform is a linear function of frequency.

(iv)   A Straight Line

A general linear function $at + b$ may be written as a linear combination of a constant and the simplest linear function, which for our purposes is

$$x_t = t - \frac{n-1}{2}, \qquad t = 0, 1, \ldots, n-1.$$

The transform of this function is

$$J_j = \frac{1}{2i} \exp\left\{ \frac{-i(n-1)\omega_j}{2} \right\} \frac{n \sin \omega_j/2 \cos n\omega_j/2 - \cos \omega_j/2 \sin n\omega_j/2}{(\sin \omega_j/2)^2}$$

for $j \neq 0$ and $J_0 = 0$. The amplitude of the transform is the absolute value of the second factor, and for large n and small ω_j this is approximately $f(n\omega_j/2)$, where

$$f(x) = \frac{x \cos x - \sin x}{x^2}.$$

A graph of this function is shown in Figure 3.2. [Note that it is the derivative of $(\sin x)/x$.] As in the case of the step function, the decay is like x^{-1}.

It should be recalled that, if the inverse transform is evaluated for values other than $0, 1, \ldots, n-1$, the periodic extension of the original series is found. In the present case, this is the sawtooth function shown in Figure 3.3, and not a continued straight line, as might have been expected.

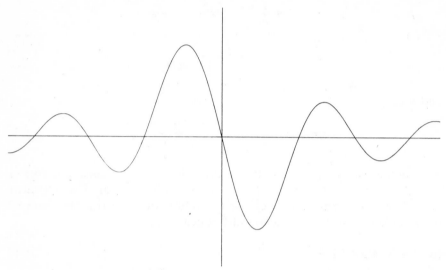

Figure 3.2   The function $(x \cos x - \sin x)/x^2$.

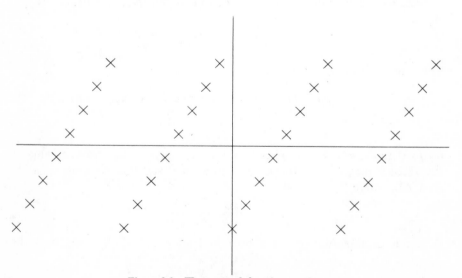

Figure 3.3   The sawtooth function, $n = 8$.

(v) Shifts, Symmetry, and Linear Phase

In all the above examples, the phase of the transform depends linearly on the index j. This is a consequence of certain symmetries in the corresponding series.

Consider first the effect of *shifting* (cyclically permuting) a series. Suppose that $y_t = x_{t+h}$ for some (integer) h, the subscripts being interpreted *modulo n*. For example, if $h = 1$ we have $y_t = x_{t+1}$, $t = 0, \ldots, n-2$, and $y_{n-1} = x_0$. The transform of the shifted series is

$$J_{y,j} = n^{-1} \sum y_t \exp(-i\omega_j t)$$

$$= n^{-1} \sum x_{t+h} \exp(-i\omega_j t)$$

$$= n^{-1} \sum x_t \exp\{-i\omega_j(t-h)\}$$

$$= n^{-1} \exp(i\omega_j h) \sum x_t \exp(-i\omega_j t)$$

$$= \exp(i\omega_j h) J_{x,j},$$

in the notation introduced at the end of Section 3.2. Thus the amplitudes of the transforms are the same, and the phases differ by a linear function of ω, the coefficient h being precisely the shift.

Now consider a series symmetric about zero (we say that such a series has *even symmetry*), that is, $x_{-t} = x_{n-t} = x_t$, $t = 1, \ldots, n-1$. Then

$$J_j = n^{-1} \sum_t x_t \exp(-i\omega_j t)$$

$$= n^{-1} \sum_t x_{n-t} \exp(-i\omega_j t)$$

$$= n^{-1} \sum_t x_{n-t} \exp\{i\omega_j(n-t)\}$$

$$= n^{-1} \sum_t x_t \exp(i\omega_j t)$$

$$= J_j^*,$$

and hence the transform is real. The converse is also true. For an antisymmetric series (a series with *odd symmetry*), one for which $x_{-t} = -x_t$, we find that the transform is purely imaginary, and conversely.

In example (iii), the step function, the amplitude is a symmetric function, and the phase is $-\omega_j(m-1)/2$. If the series were shifted by $-(m-1)/2$, the transform would be entirely real and would thus be symmetric. We conclude that the series has even symmetry about $t=(m-1)/2$. Strictly, this argument is valid only for odd m, since the shift must be an integer. However, the conclusion is valid for even m, since a series may be symmetric either about an integer time or an odd half-integer time.

In example (iv), a shift of $-(n-1)/2$ [or of $n-(n-1)/2=(n+1)/2$] makes the transform purely imaginary, and hence this series has odd symmetry about $t=(n-1)/2$.

(vi)   Periodic Series

Suppose that the series $x_0, x_1, \ldots, x_{n-1}$ is periodic with period h, an integer, and that n is an integer multiple of h, say kh. If we examine only the first full cycle, $x_0, \ldots, x_{h-1}$, we may compute a Fourier transform and invert it, obtaining a representation of the form

$$x_t = \sum_{l=0}^{h-1} K_l \exp\left(i\frac{2\pi l}{h}t\right), \qquad t=0,\ldots,h-1.$$

Since this formula defines a periodic extension when evaluated for other t, and the original series was periodic, it follows that the formula holds for $h \leq t < n$. Also

$$\frac{2\pi l}{h} = \frac{2\pi l k}{n} = \omega_{lk},$$

and thus this formula is the inverse transform for the whole series. Hence the transform of the whole series is

$$J_j = \begin{cases} K_l & \text{if } j=kl, \\ 0 & \text{otherwise.} \end{cases}$$

In other words, the transform vanishes except at the frequency $\omega_k = 2\pi/h$ and its multiples. This frequency is called the *fundamental*, and its multiples are *harmonics*.

Note that we can guarantee, not that all of these terms are nonzero, only that the others are zero. For instance, if our periodic series were $\cos 2\pi t/h = \cos t\omega_k$, only the fundamental would be nonzero, and all the harmonics would vanish.

Exercise 3.7   A Quadratic

Find the transform of the series

$$x_t = \left(t - \frac{n-1}{2}\right)^2,$$

and an approximation for large n, small ω, of the form

$$J_j \cong \exp\left\{\frac{-i(n-1)\omega_j}{2}\right\} f\left(\frac{n\omega_j}{2}\right).$$

Should $f(x)$ be real or imaginary? At what rates does $f(x)$ decay for large x?

Exercise 3.8   Triangle Wave

The series

$$x_t = \begin{cases} t, & 0 \le t \le n/2 \\ n-t, & n/2 \le t < n \end{cases}$$

defines a triangle function that extends periodically to be a triangle wave. Find its transform and an approximation as in Exercise 3.7. Find the rate of decay of the approximation.

Exercise 3.9   A Truncated Sinusoid

The step function of example (iii) is a special case of a truncated sinusoid,

$$x_t = \begin{cases} \cos(\omega t + \phi), & 0 \le t < m, \\ 0, & m \le t < n. \end{cases}$$

Find the transform of this series, and note that in general all of its terms are nonzero, even if the frequency ω is a Fourier frequency.

3.5   SMOOTH FUNCTIONS

Consider a sequence $x_0, \ldots, x_{n-1}$ for which the cosine and sine coefficients are large only for low frequencies. This means that the sequence can be represented as a sum of terms each of which is smooth, and hence the sequence itself will be smooth. Since the sines and cosines of higher frequencies oscillate rapidly, *no* smooth sequence can contain them with a large amplitude. We conclude that a function is smooth if and only if its discrete Fourier transform is small except at low frequencies.

This qualitative argument may be made more precise in several ways. The simplest is to define the (circular) *roughness coefficient* of the data to be

$$\frac{\sum\limits_{t=0}^{n-1} (x_t - x_{t-1})^2}{\sum\limits_{t=0}^{n-1} x_t^2},$$

where x_{-1} is replaced by x_{n-1}. This reflects the idea that in a smooth sequence of points the differences between successive values are all small, and hence the numerator will be small. The denominator is essentially a scaling factor, since the roughness of a set of data is a relative quantity and should not depend on the magnitude of the numbers. Often we would also wish the coefficient not to depend on the *level* of the data (that is, to be unaffected by adding the same constant to each x_t, $t = 0, \ldots, n-1$). This could be achieved by replacing x_t by $x_t - \bar{x}$, which has very little effect on what follows.

This roughness coefficient was defined (although not by this name) in a paper by Ernst Abbe in 1863 (see Kendall, 1971). The noncircular version,

$$\frac{\sum\limits_{t=1}^{n-1} (x_t - x_{t-1})^2}{\sum\limits_{t=1}^{n-1} (x_t - \bar{x})^2},$$

is also known as the von Neumann ratio and the Durbin-Watson statistic. It has been studied by von Neumann, Kent, Bellinson, and Hart (1941), von Neumann (1941, 1942), and Hart and von Neumann (1942), and by Durbin and Watson (1950, 1951, 1971). In each case the main aim was to find the probability distribution of the ratio when the series $\{x_t\}$ has some Gaussian distribution. However, we are more interested in the coefficient as a general indicator of the roughness (or smoothness) of a series.

The noncircular definition is the more natural one, since we would not normally wish to penalize an otherwise smooth series because its end values are different. However, the circular definition is more convenient, and the difference is slight, at least for large n.

We use the complex version of the discrete Fourier transform for convenience. Then

$$x_t = \sum_{j=0}^{n-1} J_j \exp(i\omega_j t),$$

and hence

$$x_t - x_{t-1} = \sum_{j=0}^{n-1} J_j \left[\exp(i\omega_j t) - \exp\{i\omega_j(t-1)\} \right]$$

$$= \sum_{j=0}^{n-1} J_j \exp(i\omega_j t)\{1 - \exp(-i\omega_j)\}.$$

The results of Section 3.3 allow us to write the roughness coefficient as

$$\frac{\displaystyle\sum_{j=0}^{n-1} |1 - \exp(-i\omega_j)|^2 |J_j|^2}{\displaystyle\sum_{j=0}^{n-1} |J_j|^2} = \frac{2\displaystyle\sum_{j=0}^{n-1}(1 - \cos\omega_j)|J_j|^2}{\displaystyle\sum_{j=0}^{n-1} |J_j|^2}$$

$$= \frac{4\displaystyle\sum_{j=0}^{n-1}(\sin\omega_j/2)^2 |J_j|^2}{\displaystyle\sum_{j=0}^{n-1} |J_j|^2}.$$

Since $\sin\omega/2$ increases from 0 to 1 as ω increases from 0 to π, it is clear that for a fixed value of the denominator the roughness is reduced by increasing low-frequency terms at the expense of high-frequency terms.

The smoothest function in this sense is the constant function, for which the roughness coefficient vanishes. The roughest functions are those for which the transform vanishes except at the frequency or frequencies closest to π, that is,

$$x_t = A\cos\pi t,$$

$$= A(-1)^t, \qquad A \neq 0, \quad t = 0,\dots,n-1$$

if n is even, and

$$x_t = A\cos\left\{\pi\left(1 - \frac{1}{n}\right)t + \phi\right\}$$

$$= A(-1)^t\cos\left(\frac{\pi t}{n} - \phi\right), \qquad A \neq 0, \quad t = 0,\dots,n-1$$

if n is odd. In each case the series oscillates rapidly. The cosine term in the second case (which completes one half of a cycle in the span of the series) is present because of the circularity of the definition of the roughness coefficient.

If x_t is replaced by $x_t - \bar{x}$ in the definition of the coefficient, the only modification is that J_0 vanishes. The roughness coefficient is thus undefined for the constant function, and the smoothest functions are of the form $A \cos(2\pi t / n + \phi)$, which executes one complete cycle in the span of the data.

Exercise 3.10   A Smooth Sequence

Suppose that

$$x_t = f\left(\frac{t + \frac{1}{2}}{n}\right),$$

where $f(x)$ is a continuous function with a continuous derivative on the interval $(0, 1)$. Then, for large n, $\{x_t\}$ is certainly a smooth sequence. Show that the roughness coefficient is approximately

$$\frac{n^{-2} \int_0^1 f'(x)^2 \, dx}{\int_0^1 f(x)^2 \, dx}.$$

Exercise 3.11   Segments of a Series

Suppose that a series is divided into two segments, which either start with the same value or end with the same value. Show that the roughness coefficient of the whole series lies between the coefficients of the two segments.

4

THE FAST FOURIER TRANSFORM

Despite the similarity of the names, the fast Fourier transform is neither a variant of nor an alternative to the discrete Fourier transform described in Chapter 3. It is an algorithm (or rather a related class of algorithms) for *computing* the discrete Fourier transform of a data series at all of the Fourier frequencies, using relatively few arithmetic operations.

The background of the fast Fourier transform and its computational advantages are described in Section 4.1. The simplest case, in which the series length factorizes into two factors, is discussed in Section 4.2. The reader who wishes to find the basic idea of the fast Fourier transform but is not interested in details may omit Sections 4.3 and 4.4, which contain the general theory and details for computer programming, respectively. Section 4.5 describes how the algorithm may be used to carry out harmonic analysis of data series.

4.1 THE COMPUTATIONAL COST OF FOURIER TRANSFORMS

The simplest way to compute the discrete Fourier transform of a set of data $x_0, \ldots, x_{n-1}$ of length n is to evaluate the sums

$$nJ_j = \sum_{t=0}^{n-1} x_t \exp(-i\omega_j t), \qquad j = 0, \ldots, n-1, \tag{1}$$

61

in turn. For convenience we shall use the complex formulation, and we shall count the computational cost of an algorithm in terms of complex addition and multiplication. Each of these sums requires $n-1$ multiplications and $n-1$ additions, if we use the fact that the first term in each sum is just x_0. Thus the total cost is $n(n-1)$ each of additions and multiplications. We also need the value of

$$\exp(-i\omega_j t) = \exp\left\{-2\pi i\left(\frac{jt}{n}\right)\right\} \tag{2}$$

for all j and t in the range $0, 1, \ldots, n-1$. Since (2) depends only on the residue of jt *modulo* n, there are only n distinct values that could be tabulated to avoid repeated calculation.

The fast Fourier transform algorithm gives the same values (1), but the number of additions and multiplications used is of the order of $n \log_2 n$. In comparison with the simple method, the cost is reduced by a factor of the order of $(\log_2 n)/n$. For the relatively short series length of 1024, Gentleman and Sande (1966) found that the execution times for ALGOL programs using a refinement of the simple method and a fast Fourier transform were 59.1 seconds and 2.0 seconds, respectively (on an IBM 7094 computer). The advantage of the fast Fourier transform increases with n and makes the transformation of very long series feasible.

The most frequently used algorithm, described in Sections 4.2 and 4.3, was first discussed in detail by Cooley and Tukey (1965), although the basic idea was known much earlier (for a history of the development of the algorithm, see Cooley, Lewis, and Welch, 1967). An important variation (the Sande-Tukey algorithm) was introduced by Gentleman and Sande (1966). Both of these variations depend on the series length n having many small factors (being *highly composite*). A simple version of each occurs when n is a power of 2. A different algorithm described by Good (1958, 1971) requires that n possess mutually prime factors and thus cannot be used for this case. Bluestein gives an algorithm that does not depend on the factorization of n (see Brigham, 1974, p. 195).

4.2 THE TWO-FACTOR CASE

The basic idea of the fast Fourier transform can be s(two-factor case $n = n_1 n_2$, when the Cooley-Tukey and S rithms are in fact the same. For any t in the range $0, 1$ find t_1 and t_2 such that

$$t = t_1 n_2 + t_2, \qquad 0 \leqslant t_1 < n_1, \quad 0 \leqslant t_2 < n_2.$$

Specifically, t_1 is the integer part of t/n_2, and t_2 is the remainder. The numbers t_1 and t_2 are the characters (generalized digits) of t in the finite mixed-radix arithmetic with radices n_1 and n_2. Similarly, for any such t_1 and t_2, the value $t_1 n_2 + t_2$ lies in the range $0, 1, \ldots, n-1$. Hence we can associate each x_t with the corresponding pair (t_1, t_2). If we write the data in a table with n_1 rows and n_2 columns, filling the rows of the table consecutively, x_t will fall in the $(t_1 + 1)$st row and the $(t_2 + 1)$st column, which we shall call the (t_1, t_2) position:

$$
\begin{array}{cccc}
& & \xrightarrow{} & \\
x_0 & x_1 & \cdots & x_{n_2 - 1} \\
x_{n_2} & x_{n_2 + 1} & \cdots & x_{2n_2 - 1} \\
\vdots & & & \\
x_{n - n_2} & x_{n - n_2 + 1} & \cdots & x_{n - 1}
\end{array}
$$

We shall denote this entry by $y(t_1, t_2)$, so that

$$ y(t_1, t_2) = x_{t_1 n_2 + t_2}. $$

We use the reversed arithmetic with radices (n_2, n_1) to represent the index j of the transform. Specifically, we write j as $j_2 n_1 + j_1$, where $0 \leqslant j_2 < n_2$ and $0 \leqslant j_1 < n_1$.

For any integer α let $W_\alpha = \exp(-2\pi i/\alpha)$. Then the transform may be written as

$$ nJ_j = \sum_{t=0}^{n-1} x_t W_n^{jt} $$

$$ = \sum_{t_1=0}^{n_1-1} \sum_{t_2=0}^{n_2-1} x_{t_1 n_2 + t_2} W_n^{(j_2 n_1 + j_1)(t_1 n_2 + t_2)} $$

$$ = \sum_{t_1} \sum_{t_2} y(t_1, t_2) W_n^{j_2 n_1 t_2 + j_1 t_1 n_2 + j_1 t_2}, $$

since

$$ W_n^{j_2 n_1 t_1 n_2} = W_n^{j_2 t_1 n} = 1. $$

The sums may be rearranged as

$$ nJ_j = \sum_{t_2} W_n^{j_2 n_1 t_2} W_n^{j_1 t_2} \sum_{t_1} y(t_1, t_2) W_n^{j_1 t_1 n_2} $$

$$ = \sum_{t_2} W_{n_2}^{j_2 t_2} \left\{ W_n^{j_1 t_2} \sum_{t_1} y(t_1, t_2) W_{n_1}^{j_1 t_1} \right\}, $$

since

$$W_n^{n_1} = W_{n_2} \quad \text{and} \quad W_n^{n_2} = W_{n_1}.$$

Now

$$z(j_1, t_2) = \sum_{t_1} y(t_1, t_2) W_n^{j_1 t_2}$$

is the j_1th term in the transform of the n_1 numbers $y(0, t_2)$, $y(1, t_2)$,...,$y(n_1 - 1, t_2)$. Similarly

$$nJ_j = \sum_{t_2} W_{n_2}^{j_2 t_2} \{ W_n^{j_1 t_2} z(j_1, t_2) \}$$

is the j_2th term in the transform of the n_2 numbers $W_n^0 z(j_1, 0)$, $W_n^{j_1} z(j_1, 1)$,...,$W_n^{j_1(n_2-1)} z(j_1, n_2 - 1)$. Thus the overall transform of the series of length $n = n_1 n_2$ can be accomplished by a number of transforms of subseries of lengths n_1 and n_2, together with multiplication by the intermediate "twiddle factor" $W_n^{j_1 t_2}$.

The final transform nJ_j can similarly be arranged in a table with n_1 rows and n_2 columns. In computer programs, it is usually the *same* table as the data were stored in at the start of the process. However, since the mixed radix representation of j used n_1 and n_2 in the reverse order, consecutive entries in the transform fall in the same *column*:

$$
\begin{array}{cccc}
J_0 & J_{n_1} & \cdots & J_{n-n_1} \\
J_1 & J_{n_1+1} & \cdots & J_{n-n_1+1} \\
\downarrow \quad \vdots & \vdots & & \vdots \\
J_{n_1-1} & J_{2n_1-1} & \cdots & J_{n-1}
\end{array}
$$

Hence the transform has to be copied out in a different order, a process known as "unscrambling." For more details, see Section 4.4.

In computing the transform for all of the Fourier frequencies, we have to carry out n_2 transforms of length n_1 and n_1 transforms of length n_2, as well as $n = n_1 n_2$ twiddle factor multiplications. If we use the simple evaluation for these shorter transforms, the total computational cost is $n_2 n_1 (n_1 - 1) + n_1 n_2 (n_2 - 1) = n_1 n_2 (n_1 + n_2 - 1)$ additions and $n_1 n_2 (n_1 + n_2 - 1) + n_1 n_2 = n_1 n_2 (n_1 + n_2)$ multiplications. These numbers may be compared with $n(n-1) = n_1 n_2 (n_1 n_2 - 1)$, the number of additions and multiplications used in the simple method. Both costs are reduced roughly by the factor $(n_1 + n_2)/n_1 n_2$. For instance, if $n_1 = n_2 = 5$ the factor is 0.4, and if $n_1 = n_2 = 10$ the factor is 0.2.

In the general algorithm, described in the next section, it is assumed that n may be factorized further as $n_1 n_2 \ldots n_k$. In this case the computational cost is reduced roughly by the factor $(n_1 + n_2 + \cdots + n_k)/n$. If each factor is 2, this becomes $2k/n = (2\log_2 n)/n$, which is the result usually quoted. Bergland (1968) gives precise formulas for the number of real additions and multiplications required by some different versions of the algorithm.

Exercise 4.1 *Computational Cost of Fourier Transforms*

Suppose that we compute the discrete Fourier transforms of a complex series of length n using only *real* arithmetic, and handling real and imaginary parts separately.

(i) How many operations (real multiplications and additions) are needed to transform a series of length n? (Find an upper bound.)

(ii) How many operations are required to transform series of lengths 2, 3, 4, and 8? (NOTE: By collecting terms together you can avoid some multiplications.)

4.3 GENERAL THEORY

The difference between the Cooley-Tukey and Sande-Tukey algorithms emerges when we consider the general case $n = n_1 n_2 \ldots n_k$. We use the mixed radix representations

$$t = t_1 n_2 n_3 \ldots n_k + t_2 n_3 \ldots n_k + \cdots + t_k$$

$$= \sum_{l=1}^{k} t_l p_l \tag{3}$$

and

$$j = j_k n_{k-1} n_{k-2} \ldots n_1 + j_{k-1} n_{k-2} \ldots n_1 + \cdots + j_1$$

$$= \sum_{l=1}^{k} j_l q_l, \tag{4}$$

where $0 \leqslant t_l < n_l$, $0 \leqslant j_l < n_l$, $l = 1, \ldots, k$, and

$$p_l = \prod_{m=l+1}^{k} n_m, \qquad q_l = \prod_{m=1}^{l-1} n_m.$$

(We use the convention that the product is 1 if the upper limit is less than the lower limit.) If $y(t_1, \ldots, t_k) = x_t$, where t is determined by (3), the transform of $x_0, \ldots, x_{n-1}$ may be written as

$$nJ_j = \sum_{t=0}^{n-1} x_t W_n^{jt}$$

$$= \sum_{t_1=0}^{n_1-1} \cdots \sum_{t_k=0}^{n_k-1} y(t_1, \ldots, t_k) W_n^{jt}. \tag{5}$$

Now

$$jt = \sum_{l=1}^{k} t_l p_l \sum_{m=1}^{k} j_m q_m,$$

and we note that $p_l q_m$ is a multiple of n whenever $l < m$. Thus

$$jt \equiv \sum_{l=1}^{k} t_l p_l j_l q_l + \sum t_l p_l j_m q_m \,(mod\ n),$$

where the second sum is over the range

$$1 \leqslant l \leqslant k, \quad 1 \leqslant m \leqslant k, \quad m < l.$$

The second sum may thus be written as

$$\sum_{l=1}^{k} \sum_{m=1}^{l-1} t_l p_l j_m q_m = \sum_{l=1}^{k-1} \sum_{m=1}^{l} t_{l+1} p_{l+1} j_m q_m \tag{6}$$

or

$$\sum_{m=1}^{k} \sum_{l=m+1}^{k} t_l p_l j_m q_m = \sum_{l=1}^{k} \sum_{m=l+1}^{k} t_m p_m j_l q_l. \tag{7}$$

Thus (5) may be expanded as

$$nJ_j = \sum_{t_1} \cdots \sum_{t_k} y(t_1, \ldots, t_k) W_n^{t_k p_k j_k q_k} \prod_{l=1}^{k-1} W_n^{t_l p_l j_l q_l} \prod_{m=1}^{l} W_n^{t_{l+1} p_{l+1} j_m q_m} \tag{8}$$

$$= \sum_{t_1} \cdots \sum_{t_k} y(t_1, \ldots, t_k) \prod_{l=1}^{k} W_n^{t_l p_l j_l q_l} \prod_{m=l+1}^{k} W_n^{t_m p_m j_l q_l}. \tag{9}$$

If we adopt the convention that $t_{k+1} = 0$, expansion (8) may be written recursively as

$$y^{(0)}(t_1,\ldots,t_k)=y(t_1,\ldots,t_k)=x_t,$$

$$y^{(l)}(j_1,\ldots,j_{l-1},j_l,t_{l+1},\ldots,t_k)$$

$$=\prod_{m=1}^{l} W_n^{t_{l+1}p_{l+1}j_mq_m} \sum_{t_l=0}^{n_l-1} y^{(l-1)}(j_1,\ldots,j_{l-1},t_l,t_{l+1},\ldots,t_k)\, W_{n_l}^{t_lj_l},$$

$$l=1,\ldots,k,\quad(10)$$

and

$$nJ_j=y^{(k)}(j_1,\ldots,j_k),$$

where j is given by (4). Expansion (9), which is algebraically identical to (8), gives rise to an alternative recursive representation:

$$z^{(0)}(t_1,\ldots,t_k)=y(t_1,\ldots,t_k)=x_t,$$

$$z^{(l)}(j_1,\ldots,j_{l-1},j_l,t_{l+1},\ldots,t_k)$$

$$=\prod_{m=l+1}^{k} W_n^{t_m p_m j_l q_l} \sum_{t_l=0}^{n_l-1} z^{(l-1)}(j_1,\ldots,j_{l-1},t_l,t_{l+1},\ldots,t_k)\, W_{n_l}^{t_lj_l},$$

$$l=1,\ldots,k,\quad(11)$$

and

$$nJ_j=z^{(k)}(j_1,\ldots,j_k),$$

where again j is given by (2).

Equation (10) is the twiddle factor form of the Cooley-Tukey algorithm, given by Gentleman and Sande (1966). The original Cooley-Tukey algorithm (Cooley and Tukey, 1965) is a similar recursion, found by using the left-hand side of (6) instead of the right-hand side. Equation (11) is the Sande-Tukey algorithm, also given by Gentleman and Sande (1966), which exists only in a twiddle factor form.

In both algorithms the t indices are progressively "transformed out" and replaced by j indices. In each case, at the lth step we have to transform the

$n/n_l = p_l q_l$ series, each of length n_l, and multiply the result by a twiddle factor. The algorithms differ only in the way in which the twiddle factor is calculated. In (10), the twiddle factor may be rewritten as

$$W_{q_{l+2}}^{\alpha t_{l+1}}$$

where

$$\alpha = \sum_{m=1}^{l} j_m q_m, \tag{12}$$

whereas in (11) the twiddle factor is

$$W_{p_{l-1}}^{\beta j_l},$$

where

$$\beta = \sum_{m=l+1}^{k} t_m p_m. \tag{13}$$

The values α and β are the low-order terms in the expansions of j and t, respectively. This apparently minor difference becomes important when the algorithms are implemented on a computer (see the next section).

Exercise 4.2   The Good Algorithm (Good, 1958, 1971)

If n has k *pairwise mutually prime* factors $n_1, \ldots, n_k$ (that is, no pair of factors has a common factor), the residues *modulo n* of

$$\sum_{l=1}^{k} \frac{t_l n}{n_l}$$

as the t_l vary over the domain $0 \leqslant t_l < n_l$, $l = 1, \ldots, k$, are precisely the values $0, 1, \ldots, n-1$. Thus any t in this range may be written uniquely as

$$t \equiv \sum_{l=1}^{k} \frac{t_l n}{n_l} (mod\ n).$$

The identification $y(t_1, \ldots, t_k) = x_t$, where t is given by this expression, provides another way of indexing the data as a k-way array. The transform may now be written as

$$nJ_j = \sum_t x_t W_n^{jt} = \sum_{t_1} \cdots \sum_{t_k} y(t_1, \ldots, t_k) W_n^{jt}$$

and

$$W_n^{jt} = \Pi \, W_n^{jt_l n / n_l}$$

$$= \Pi \, W_{n_l}^{jt_l}$$

$$= \Pi \, W_{n_l}^{j_l t_l},$$

where j_l is the residue of j *modulo* n_l. Use this expansion to obtain a recursive algorithm for computing nJ_j. Note that each step involves discrete Fourier transforms of subseries, and that there are no twiddle factors.

What is the major problem in unscrambling the transform?

Exercise 4.3   *The Cooley-Tukey Algorithm (Cooley and Tukey, 1965)*

The original (untwiddled) Cooley-Tukey algorithm may be derived as follows. Take the expansion of *jt* [below (5)] and omit all terms that are multiples of *n*. Write the remaining terms as

$$\sum_{l=1}^{k} t_l p_l r_l,$$

where r_l is a sum of j's and q's. Factorize W_n^{jt} as $\Pi W_n^{t_l p_l r_l}$, and expand (5) accordingly. Finally write the multiple sum in an appropriate recursive form. Note that the steps of the recursion are no longer discrete Fourier transforms of subseries.

4.4   PROGRAMMING CONSIDERATIONS

The fast Fourier transform is used in many different ways on many different computers. The way it is implemented for any given purpose depends on the type of data to which it will be applied, the computer that is to be used, and the numerical precision needed, amongst other things. For instance, if it is always to be used on series of the same length, a less general program is needed, which can therefore be made somewhat more efficient. Sometimes special hardware is available which either carries out the whole transform or may be used for the elementary transforms. However, in most situations there is no preferred series length and no special hardware, and a general purpose program is called for.

Since it may be required to transform very long series, a general purpose fast Fourier transform program should use only a fixed amount of internal storage. Hence the transformed series should be computed in the section of memory in which the original series was passed to it, that is, the algorithm should carry out the transform *in place*. Each of the elementary transforms is computed from a number of (complex) values, and the same number of values is computed from them.

If we restrict the size of the factors n_l, say $n_l \leqslant N$, then an in-place algorithm can be derived as follows. An array of length N is allocated internally and is used to compute the elementary transforms in turn. As each transform is completed, it is copied back in place of the values that have just been transformed.

There are other advantages to limiting the factors that will be used. For instance, if $N = 2$, that is, the series length is restricted to be a power of 2, all of the elementary transforms are of subseries of length 2. But the transform of two values consists of the sum and the difference of the numbers, and thus requires no multiplications. Similarly the transform of a series of length 4 can be found without multiplications.

One disadvantage of in-place algorithms is caused by the mixed radix representations (3) and (4) of t, the time index, and j, the frequency index, respectively. The first step of either the Cooley-Tukey algorithm (10) or the Sande-Tukey algorithm (11) is to change the indexing from a single-dimensioned vector of length n to an array with k indices. This can be done without moving the data if we regard the tth cell of the vector as the $(t_1, \ldots, t_k)$th cell of the array, where t and $t_1, \ldots, t_k$ are related by (3). However, this means that at the end of the recursion we have $y^{(k)}(j_1, \ldots, j_k)$ or $z^{(k)}(j_1, \ldots, j_k)$ stored in the $(j_1 n_2 \ldots n_k + \cdots + j_k)$th cell of the vector, whereas the frequency index of this term is $j = j_k n_{k-1} \ldots n_1 + \cdots + j_1$. In other words, the transform is stored in *digit-reversed* sequence and usually has to be "unscrambled" before it is returned.

Alternatively, we may identify the cells of the vector with the cells of the array by using relation (4). In this case the transform is computed in its correct sequence, but the data have to be "scrambled" to initialize the recursion.

Suppose that we decide to leave the data in place and to unscramble the transform. The programming of either (10) or (11) is now relatively straightforward. There is an outer loop in which l is incremented from 1 to k. For fixed values of $j_1, \ldots, j_{l-1}$ and $t_{l+1}, \ldots, t_k$, we have to transform a subseries of length n_l. To find this subseries in the vector, we need only find the first value, and then take every p_lth value until we have all n_l values. The index of the first value is

$$(j_1 n_2 \ldots n_{l-1} n_l \ldots n_k + \cdots + j_{l-1} n_l \ldots n_k) + (t_{l+1} n_{l+2} \ldots n_k + \cdots + t_k)$$

$$= (j_1 n_2 \ldots n_{l-1} + \cdots + j_{l-1}) n_l \ldots n_k + (t_{l+1} n_{l+2} \ldots n_k + \cdots + t_k).$$

$$(14)$$

As $j_1,\ldots,j_{l-1}$ are varied, the first term takes as values all multiples of $n_l\ldots n_k = p_{l-1}$ from 0 up to (but not including) the $n_1\ldots n_{l-1}$th, or the q_lth. Similarly, as $t_{l+1},\ldots,t_k$ are varied, the second term takes on all values from 0 to $n_{l+1}\ldots n_k - 1 = p_l - 1$. Thus we can carry out the whole of the lth step by incrementing these two terms explicitly, instead of incrementing the t's and j's individually.

At this point the Sande-Tukey algorithm (11) has a clear advantage, for the computation of the twiddle factor in (11) involves only j_l and β, defined in (13). But β is precisely the second term in (14) and has therefore already been calculated. In other words, for the Sande-Tukey algorithm we need, not the values of $j_1,\ldots,j_{l-1}$ or $t_{l+1},\ldots,t_k$ individually, only the values of the two terms of (14). By contrast, in the Cooley-Tukey algorithm we need the value of α in (12), for which we require the values of $j_1,\ldots,j_{l-1}$ individually.

The situation is reversed if we decide to scramble the data and compute the transform in sequence. It is now the Cooley-Tukey algorithm that gives the simpler indexing. Neither algorithm has any inherent advantage, except in situations where either the data are available in scrambled order, or the transform may be used without unscrambling (see, for instance, Section 7.4).

Our discussion so far has been entirely in terms of complex data, since the algorithm may be described (and programmed) more compactly. Furthermore, a program written to analyze complex data may be used with real data by supplying a zero imaginary part. However, there is some redundancy in this, and a more efficient algorithm may be obtained by a slight modification.

If we have *two* real series $\{x_t\}$ and $\{y_t\}$ to be analyzed, the simplest and most efficient procedure is to construct a fictitious complex series $z_t = x_t + iy_t$. It is easily verified that

$$J_{x,j} = \frac{J_{z,j} + J_{z,n-j}^*}{2},$$

$$J_{y,j} = \frac{J_{z,j} - J_{z,n-j}^*}{2i},$$

and hence the two transforms may be disentangled easily from the transform of $\{z_t\}$. If we have only a single series $\{x_t\}$ of even length $n = 2m$, say, we construct the series

$$z_t = x_{2t} + ix_{2t+1}, \qquad t = 0,\ldots,m.$$

It may then be verified that

$$J_{x,j} = \frac{J_{z,j} + J_{z,m-j}^*}{2} + \frac{\exp(-i\omega_j)(J_{z,j} - J_{j,m-j}^*)}{2i}.$$

This involves a little more calculation, but still gives a substantial saving.

Exercise 4.4   Character Reversal

Suppose that t and j are in the range $0, 1, \ldots, n-1$, and have the mixed radix representations (3) and (4), respectively. Suppose in addition that $t_l = j_l$, $l = 1, \ldots, k$. Describe an algorithm for obtaining the value of j from that of t. (NOTE: This is the unscrambling problem.)

Exercise 4.5   Unscrambling: Radix 2

In the radix-2 case, where each factor is restricted to be 2, the mixed radix representation is the binary expansion. Write down the order in which the transform is computed for (i) $n=4$, (ii) $n=8$, (iii) $n=16$. Note that unscrambling consists only of pairwise interchanges.

Exercise 4.6   Unscrambling: General Case

Show that the unscrambling operation can be achieved by pairwise interchanges if and only if the factors $n_1, \ldots, n_k$ satisfy $n_l = n_{k-l+1}$, $l = 1, \ldots, k$.

4.5   APPLICATION TO HARMONIC ANALYSIS OF DATA

The efficiency of the fast Fourier transform algorithm has made possible the routine harmonic analysis of extensive sets of data. However, the gains are dramatic only when the series length n is highly composite. Although this can often be arranged, some awkward numbers arise quite frequently. For instance, 365 factorizes into 5×73, so that a whole number of years of daily data can be a problem. Sometimes a small part of the data can be discarded to have a more convenient number. However, this is often unsatisfactory, since it inevitably involves some loss of information.

There is another solution if we are prepared to have the Fourier transform calculated at a different set of frequencies. The transform may be defined as a function of a continuous frequency variable ω, as

$$J(\omega) = n^{-1} \sum x_t \exp(-it\omega), \qquad -\pi \leqslant \omega \leqslant \pi \quad \text{or} \quad 0 \leqslant \omega \leqslant 2\pi.$$

The Fourier frequencies may then be regarded just as a convenient grid at which to evaluate this function, if we are not concerned with the orthogonality properties described in Chapter 3. It may be evaluated at a more finely spaced grid as follows.

Suppose that we have data $x_0, \ldots, x_{n-1}$, and that $n' > n$. In practice we shall choose n' to be convenient for the fast Fourier transform, and in order to use the program given in the Appendix to this chapter n' must be a power of 2. We now extend the data by zeros; let

$$x'_t = \begin{cases} x_t, & 0 \leqslant t < n \\ 0, & n \leqslant t < n' \end{cases}$$

be the extended data. The transform of the extended series is

$$J'_j = \frac{1}{n'} \sum_{t=0}^{n'-1} x'_t \exp(-i\omega'_j t)$$

$$= \frac{1}{n'} \sum_{t=0}^{n-1} x_t \exp(-i\omega'_j t)$$

$$= \frac{n}{n'} J(\omega'_j),$$

where $\omega'_j = 2\pi j / n'$ is the jth Fourier frequency for series length n'. We thus obtain (a multiple of) the required transform, but at the Fourier frequencies ω'_j, whose spacing is $2\pi/n' < 2\pi/n$.

The inversion formula allows us to retrieve all the original data, and hence no information has been lost. However, none has been gained either, for the inversion formula also implies that

$$\sum_{j=0}^{n'-1} J'_j \exp(i\omega'_j t) = 0, \qquad n \leqslant t < n'.$$

In other words, the computed transform satisfies $n' - n$ constraints.

The extension of data by zeros can be beneficial in certain applications, such as the computation of *autocovariances* (see Section 7.4). However, it emphasizes the problem of *leakage*: the phenomenon in which the presence of a particular harmonic component causes the transform to be nonzero at other frequencies. Example (i) of Section 3.4 showed that any frequency other than a Fourier frequency causes leakage into all Fourier frequencies. In the present case, however, a sinusoid in the original data becomes a truncated sinusoid in the extended data, and it was shown in Exercise 3.9 that the transform of this is in general nonzero at the Fourier frequencies, for *any* frequency in the original data. Thus *any* harmonic component in the original data will give rise to leakage in the transform of the extended data. This makes the control of leakage more obviously desirable, although it is in fact essential even in the analysis of nonextended data. Techniques for leakage reduction are discussed in Section 5.2.

Exercise 4.7   Correcting for the Mean

Suppose that the series $x_0, \ldots, x_{n-1}$ shows relatively small fluctuations y_t about a large average value, x (that is, $x_t = x + y_t$, where $|y_t| \ll |x|$). The series is analyzed by extending it with zeros to a length $n' > n$, followed by calculation of the discrete Fourier transform (at the Fourier frequencies for n').

(i)   Show that leakage from the zero-frequency component x is present in all terms of the transform. How large would x have to be for this leakage to dominate the transform of $\{y_t\}$?

(ii)   Show that, if the mean of the *extended* series is subtracted from each term in the *extended* series, this leakage is unaffected.

(iii)   Show that, if the mean of the *original* series is subtracted out *before* extension by zeros, the leakage is removed completely.

[Note that results (ii) and (iii) are true only up to the numerical accuracy of the calculations. This qualification would usually be unimportant in (iii), but may be significant in (ii).]

(iv)   Show that these considerations hold also for any periodic component in the data whose frequency is a Fourier frequency for the original, unextended data.

APPENDIX

The following program is a FORTRAN implementation of the Sande-Tukey fast Fourier transform (see Section 4.3). It is an in-place algorithm (see Section 4.4). Most of the internal storage is used for the arrays UR and UI, which contain the values of $\cos(\pi/2^j)$ and $\sin(\pi/2^j)$, $j = 1, \ldots, 15$, respectively. These values are computed recursively, using only the SQRT function. Other sines and cosines needed for twiddle factors are computed from these tables.

```
      SUBROUTINE FT01A (XR,XI,N,INV)
C
C     THIS SUBROUTINE IMPLEMENTS THE SANDE-TUKEY RADIX-2
C     FAST FOURIER TRANSFORM.  EITHER THE DIRECT OR
C     THE INVERSE TRANSFORM MAY BE COMPUTED.   PARAMETERS ARE
C
C     NAME    TYPE                        VALUE
C                              ON ENTRY                  ON RETURN
C
C     XR    REAL ARRAY REAL PART OF THE        REAL PART OF THE
C                      SERIES                  TRANSFORM
C
C     XI    REAL ARRAY IMAGINARY PART OF       IMAGINARY PART
C                      THE SERIES              OF THE TRANSFORM
C
C     N     INTEGER    THE SERIES LENGTH       UNCHANGED
C
C     INV   INTEGER    0  FOR DIRECT TRANSFORM   INV IS SET TO
C                      1  FOR INVERSE TRANSFORM  -1   TO INDICATE
C                                                ERROR RETURN
C
C     THE DIRECT TRANSFORM IS -
C
C                 N
C        (1/N) SUM X(T)*EXP(-2*PI*I*(T-1)*(J-1)/N), J=1 TO N
C              T=1
C
C     THE INVERSE TRANSFORM IS -
C
C                 N
C              SUM X(T)*EXP( 2*PI*I*(T-1)*(J-1)/N), J=1 TO N
C              T=1
C
      DIMENSION XR(N),XI(N),UR(15),UI(15)
      LOGICAL FIRST
      DATA FIRST /.TRUE./
      IF(.NOT. FIRST) GO TO 120
      UR(1)=0.0
      UI(1)=1.0
      DO 110 I=2,15
      UR(I)=SQRT((1.0+UR(I-1))/2.0)
110   UI(I)=UI(I-1)/(2.0*UR(I))
      FIRST=.FALSE.
120   IF(N .GT. 0 .AND.  N .LE. 2**16) GO TO 130
      INV=-1
      RETURN
130   NO=1
      II=0
140   NO=NO+NO
      II=II+1
      IF(NO .LT. N) GO TO 140
      I1=NO/2
      I3=1
      IO=II
```

```
      DO 260 I4=1,II
      DO 250 K=1,I1
      WR=1.0
      WI=0.0
      KK=K-1
      DO 230 I=1,I0
      IF(KK .EQ. 0) GO TO 240
      IF(MOD(KK,2) .EQ. 0) GO TO 230
      J0=I0-I
      WS=WR*UR(J0)-WI*UI(J0)
      WI=WR*UI(J0)+WI*UR(J0)
      WR=WS
  230 KK=KK/2
  240 IF(INV .EQ. 0) WI=-WI
      L=K
      DO 250 J=1,I3
      L1=L+I1
      ZR=XR(L)+XR(L1)
      ZI=XI(L)+XI(L1)
      Z      =WR*(XR(L)-XR(L1))-WI*(XI(L)-XI(L1))
      XI(L1)=WR*(XI(L)-XI(L1))+WI*(XR(L)-XR(L1))
      XR(L1)=Z
      XR(L)=ZR
      XI(L)=ZI
  250 L=L1+I1
      I0=I0-1
      I3=I3+I3
  260 I1=I1/2
      UM=1.0
      IF(INV .EQ. 0) UM=1.0/FLOAT(NO)
      DO 310 J=1,NO
      K=0
      J1=J-1
      DO 320 I=1,II
      K=2*K+MOD(J1,2)
  320 J1=J1/2
      K=K+1
      IF(K .LT. J) GO TO 310
      ZR=XR(J)
      ZI=XI(J)
      XR(J)=XR(K)*UM
      XI(J)=XI(K)*UM
      XR(K)=ZR*UM
      XI(K)=ZI*UM
  310 CONTINUE
      RETURN
      END
```

5

EXAMPLES OF HARMONIC ANALYSIS

The theory of harmonic analysis was described in Chapter 3, and a numerical algorithm for carrying it out with relatively little computational effort was given in Chapter 4. In this chapter we present some examples of harmonic analysis of data series and illustrate the insights that such analyses can (and cannot), provide about the data. We also mention in Sections 5.2 and 5.7 two ways in which the series may be modified before the analysis to make the interesting features of the data more apparent.

5.1 THE VARIABLE-STAR DATA

As the first example we consider the variable-star data (Figure 1.1), which was used in Chapter 2 to illustrate the model of hidden periodicities. In that chapter it was shown that these data consist of two strong sinusoidal components plus residuals that are of precisely the order of magnitude we would expect to arise from rounding off the data to integers. We shall now use harmonic analysis to see what structure, if any, there is to these residuals. Harmonic analysis can detect whether any behavior of interest occurs at frequencies other than the two discovered in Section 2.4.

Figure 5.1 shows the periodogram of the variable-star data. The function graphed is the base-10 logarithm of $\tilde{R}(\omega)^2 = \tilde{A}(\omega)^2 + \tilde{B}(\omega)^2$, defined in

Section 2.3, where

$$\tilde{A}(\omega) = \frac{2}{n} \sum (x_t - \bar{x}) \cos \omega t, \qquad \tilde{B}(\omega) = \frac{2}{n} \sum (x_t - \bar{x}) \sin \omega t.$$

This differs from the periodogram

$$I(\omega) = \frac{1}{2\pi n} \left[\left\{ \sum (x_t - \bar{x}) \cos \omega t \right\}^2 + \left\{ \sum (x_t - \bar{x}) \sin \omega t \right\}^2 \right]$$

by a factor of $8\pi/n$. The function $\tilde{R}(\omega)^2$ is more useful in harmonic analysis, since it is the (least squares estimate of the) squared amplitude of a sinusoid with frequency ω. The radix-2 fast Fourier transform listed in the Appendix to Chapter 4 was used to compute this function. The data were extended to a series length of $n' = 1024$ (the next power of 2 above $n = 600$, the actual series length) by appending 424 zeros, after subtracting the mean of the original series (see Exercise 4.7). The discrete Fourier transform was then calculated at the Fourier frequencies for n', namely $\omega_j' = 2\pi j/n', j = 0, \ldots, n' - 1$. The first term, $J(0)$, may be ignored, since it is just the mean of the series being analyzed, which is necessarily zero. If the transform were carried out in exact arithmetic it would be zero, and its computed value consists only of accumulated numerical errors. The last $511 (= n'/2 - 1)$ terms in the transform may also be ignored, since they are the complex conjugates of the 2nd to 512th terms, in reverse order. The graph shows the values of $\log_{10} \tilde{R}(\omega_j')^2$ for $j = 1, \ldots, 512$, that is, for $0 < \omega_j \leqslant \pi$, plotted against j.

The logarithm of the periodogram is preferable for plotting because of the variation in the order of magnitude of the periodogram between different frequencies. Had the periodogram itself been plotted, on such a scale that the largest values were on the graph, the detail at higher frequencies would have been completely obscured.

The peaks corresponding to the two components are the dominant features of the graph. The largest ordinates are at $j = 35$ and $j = 43$, respectively. The values $\omega_{35}' = 2\pi \times 35/1024 = 0.2148$ and $\omega_{43}' = 2\pi \times 43/1024 = 0.2638$ would have provided adequate starting values for the iterative estimation procedure of Section 2.4. This would be the usual procedure for determining the number of components and the starting values of their frequencies.

The moderately large values of the periodogram ordinates close to these peaks are caused by *leakage* from the peaks. We know from the results of Section 2.4 that no other components with amplitudes as large as 1.0 (corresponding to the value 0 on this graph) could be present in the data, since the data may be written as the two components, plus residuals with an average squared value of around $\frac{1}{12}$.

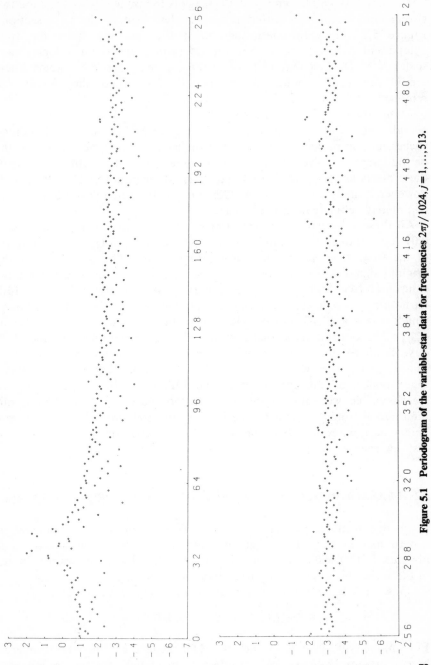

Figure 5.1 Periodogram of the variable-star data for frequencies $2\pi j/1024, j = 1, \ldots, 513$.

The simplest way to remove the leakage is to remove its sources, the two strong components. (A different way is described in the next section.) Figure 5.2 shows the periodogram of the residuals from the two-component model fitted in Section 2.4 (also graphed on a logarithmic scale). We note first that all trace of the two main peaks has gone. There are in fact slight "holes" in the periodogram where they have been removed.

Comparison of Figures 5.1 and 5.2 shows that the leakage in Figure 5.1 actually extends to *all* frequencies. In Figure 5.2 we see a well-defined succession of peaks, separated by troughs in which the values of the periodogram are lower by up to two or three orders of magnitude. Some of these peaks are also visible in Figure 5.1, although not as clearly. Others, however, namely, those near the 70th, 84th, 171st, and 177th ordinates, are completely submerged in the leakage.

All these subsidiary peaks, the largest of which have heights of around 0.01 (corresponding to amplitudes of approximately 0.1), are at multiples of the frequencies of the two strong components. This suggests that the original signal is the sum of two *periodic* terms, each of which is almost sinusoidal [see Section 3.4, example (vi)]. It is curious that the 14th harmonic of the lower frequency component falls at the 494th ordinate, which is $18 = 36/2$ short of the 512th. Thus the 29th harmonic would fall at $\omega'_{1024} = 2\pi$, confirming our suspicion that this component has a period of 29 days. Similarly, the 12th harmonic of the higher-frequency component falls exactly at $\omega'_{512} = \pi$, which corresponds to a period of precisely 24 days. Since almost all harmonics up to the 14th and 12th, respectively, are clearly visible, we would expect to see further harmonics aliased back into the interval from 0 to π. If the periods are precisely integers, however, the higher harmonics alias back on top of lower harmonics and thus are not distinguishable.

5.2 LEAKAGE REDUCTION BY DATA WINDOWS: TAPERS AND FADERS

It was demonstrated in Section 5.1 that leakage from strong components in a periodogram may be reduced by removing its source, the strong components. In this section a method is described that leaves the sources intact in the series, but reduces the magnitude of the leakage dramatically.

Consider first the simple case of data consisting of a pure sinusoid:

$$x_t = R \exp\{i(\lambda t + \phi)\}, \qquad t = 0, \ldots, n - 1.$$

For simplicity we take $R = 1$ and $\phi = 0$. [The general case may be recovered by multiplying by $R \exp(i\phi)$]. The transform of this series, regarded as a

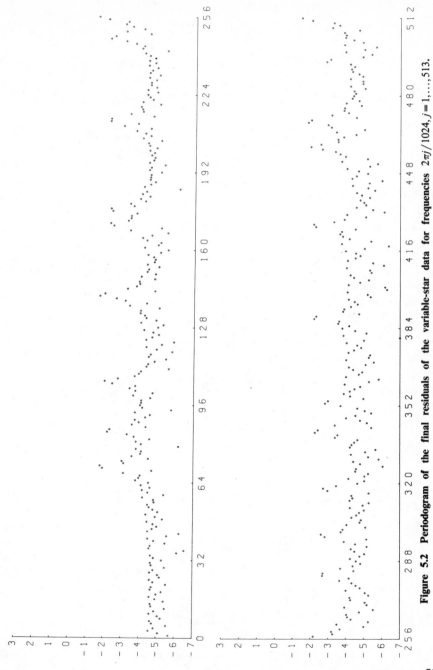

Figure 5.2 Periodogram of the final residuals of the variable-star data for frequencies $2\pi j/1024, j=1,\dots,513$.

function of a continuous variable ω, is (see Exercise 2.2.)

$$J(\omega) = n^{-1} \sum_{t=0}^{n-1} x_t \exp(-i\omega t)$$

$$= \exp\left\{\frac{i(n-1)(\lambda-\omega)}{2}\right\} \frac{\sin\{n(\lambda-\omega)/2\}}{n\sin\{(\lambda-\omega)/2\}}.$$

Let

$$K(\omega) = J(\omega)\exp\left\{\frac{i(n-1)\omega}{2}\right\}, \tag{1}$$

which in this case is therefore

$$\exp\left\{\frac{i(n-1)\lambda}{2}\right\} \frac{\sin\{n(\lambda-\omega)/2\}}{n\sin\{(\lambda-\omega)/2\}}. \tag{2}$$

Expression 1 is the transform of the data shifted (circularly) by $-(n-1)/2$, and corresponds to moving the time origin to the middle of the data [see Section 3.4, example (v)]. Now, for large n and $\omega \neq \lambda$,

$$K\left(\omega - \frac{2\pi}{n}\right) = \exp\left\{\frac{i\lambda(n-1)}{2}\right\} \frac{\sin\{n(\lambda-\omega)/2+\pi\}}{n\sin\{(\lambda-\omega)/2+\pi/n\}}$$

$$= -\exp\left\{\frac{i\lambda(n-1)}{2}\right\} \frac{\sin\{n(\lambda-\omega)/2\}}{n\sin\{(\lambda-\omega)/2+\pi/n\}}$$

$$\cong -\exp\left\{\frac{i\lambda(n-1)}{2}\right\} \frac{\sin\{n(\lambda-\omega)/2\}}{n\sin\{(\lambda-\omega)/2\}}$$

$$= -K(\omega),$$

and similarly $K(\omega+2\pi/n) \cong -K(\omega)$. If we let

$$K_H(\omega) = \tfrac{1}{4}K\left(\omega - \frac{2\pi}{n}\right) + \tfrac{1}{2}K(\omega) + \tfrac{1}{4}K\left(\omega + \frac{2\pi}{n}\right), \tag{3}$$

it follows that $K_H(\omega) \cong 0$, to this order of approximation. More precisely,

$$K_H(\omega) = \frac{1}{4n} \exp\left\{\frac{i\lambda(n-1)}{2}\right\} \sin\left\{\frac{n(\lambda-\omega)}{2}\right\}$$

$$\times \left[\frac{2}{\sin\{(\lambda-\omega)/2\}} - \frac{1}{\sin\{(\lambda-\omega)/2-\pi/n\}} - \frac{1}{\sin\{(\lambda-\omega)/2+\pi/n\}}\right].$$

If n is large, we may write

$$\sin\left\{\frac{\lambda-\omega}{2} - \frac{\pi}{n}\right\} \cong \sin\left(\frac{\lambda-\omega}{2}\right) - \frac{\pi}{n}\cos\left(\frac{\lambda-\omega}{2}\right),$$

$$\sin\left\{\frac{\lambda-\omega}{2} + \frac{\pi}{n}\right\} \cong \sin\left(\frac{\lambda-\omega}{2}\right) + \frac{\pi}{n}\cos\left(\frac{\lambda-\omega}{2}\right),$$

and hence, after some simplification,

$$K_H(\omega) \cong -\frac{\pi^2}{2n^3}\exp\left\{\frac{i\lambda(n-1)}{2}\right\}\sin\left\{\frac{n(\lambda-\omega)}{2}\right\}$$

$$\times \frac{[\cos\{(\lambda-\omega)/2\}]^2}{\sin\{(\lambda-\omega)/2\}([\sin\{(\lambda-\omega)/2\}]^2 - (\pi^2/n^2)[\cos\{(\lambda-\omega)/2\}]^2)}$$

$$\cong \frac{-\pi^2}{2n^3}\exp\left\{\frac{i\lambda(n-1)}{2}\right\}\sin\left\{\frac{n(\lambda-\omega)}{2}\right\}\frac{[\cos\{(\lambda-\omega)/2\}]^2}{[\sin\{(\lambda-\omega)/2\}]^3}. \qquad (4)$$

Operation 3 is known as *hanning*, and we use the subscript H to distinguish its results. The leakage in the hanned version of K, given approximately by (4), is better behaved than that for K itself, given exactly by (2), in two ways. First, it contains a factor n^{-3} rather than n^{-1}; this means that it is much smaller in long series. Second, the factor $[\sin\{(\lambda-\omega)/2\}]^{-3}$ decays like $|\lambda-\omega|^{-3}$ rather than $|\lambda-\omega|^{-1}$, and thus for a given value of n the leakage diminishes more rapidly as ω moves away from λ, the position of the peak. Hence the leakage is reduced in magnitude and is contained more closely around $\omega=\lambda$.

If we use (3) and (1) to define $K_H(\omega)$ for an arbitrary series $\{x_t\}$, then (see Exercise 5.1)

$$K_H(\omega) = n^{-1}\exp\left\{\frac{i\omega(n-1)}{2}\right\}\sum_{t=0}^{n-1}x_t\exp(-i\omega t)\frac{1-\cos\{2\pi(t+\frac{1}{2})/n\}}{2}.$$

$$(5)$$

Thus

$$J_H(\omega) = \exp\left\{\frac{-i\omega(n-1)}{2}\right\}K_H(\omega) \qquad (6)$$

is the transform of the series

$$y_t = \frac{x_t\left[1 - \cos\{2\pi(t+\frac{1}{2})/n\}\right]}{2}, \qquad t = 0,\ldots,n-1. \tag{7}$$

We shall call (6) the *hanned transform* of $\{x_t\}$, although it should be pointed out that this expression is *not* found from $J(\omega)$ by the analog of (3). Equation 7 shows that an alternative way to compute $J_H(\omega)$ is to multiply the data $x_0,\ldots,x_{n-1}$ by the *data window* (or *fader*)

$$w_t = \frac{1}{2}\left[1 - \cos\left\{\frac{2\pi\left(t+\frac{1}{2}\right)}{n}\right\}\right], \qquad t = 0,\ldots,n-1 \tag{8}$$

and transform the *windowed* or *tapered* data.

When we analyze data without first tapering it, we are, by default, using the rectangular data window (or *boxcar*)

$$v_t = \begin{cases} 1, & 0 \leqslant t < n, \\ 0, & \text{otherwise} \end{cases}.$$

The sequence $\{w_t\}$ may be thought of as a smooth approximation to the boxcar. Now the transform of a tapered sinusoid is just the transform of the data window centered at the frequency of the sinusoid (see Exercise 5.2). Hence leakage in the transform of a tapered sinusoid is caused by the sidelobes of the transform of the data window (which is the Dirichlet function for the boxcar). It was shown in Sections 3.4 and 3.5 that the transform of a smooth series decays more rapidly than that of a rough series. We conclude that it is the smoothness of the data window $\{w_t\}$ which leads to its good leakage characteristics.

We may therefore use any smooth approximation to the boxcar as a data window. We would usually prefer a window that leaves the bulk of the data unmodified, and just tapers the ends. A convenient window may be found by separating the two halves of the cosine bell (8) and inserting a stretch of 1's. This gives the window

$$u_t = \begin{cases} \frac{1}{2}\left[1 - \cos\left\{\frac{\pi\left(t-\frac{1}{2}\right)}{m}\right\}\right], & t = 0,\ldots,m-1, \\[2mm] 1, & t = m,\ldots,n-m-1, \\[2mm] \frac{1}{2}\left[1 - \cos\left\{\frac{\pi\left(n-t+\frac{1}{2}\right)}{m}\right\}\right], & t = n-m,\ldots,n-1, \end{cases} \tag{9}$$

where m is chosen so that $2m/n$, the proportion of the data that is tapered, is some desired value. Tukey (1967) has suggested that 10% or 20% may be suitable. It may be seen (see Exercise 5.5) that this window gives an intermediate reduction in leakage. The FORTRAN subroutine TAPER, presented in the Appendix to this chapter, uses this form of tapering. Figure 5.3 shows the boxcar and cosine bell data windows and the split cosine bell with 50% tapering. The transforms of these data windows are shown in Figure 5.4.

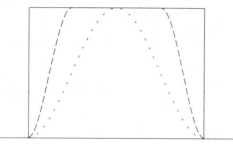

Figure 5.3   The boxcar (solid line), the cosine bell window (short dashed line), and the split cosine bell window, $p = 0.5$ (long dashed line).

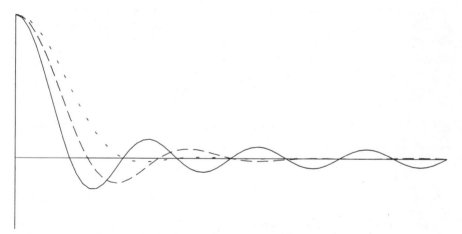

Figure 5.4   Transforms of the data windows shown in Figure 5.3 for a series of length 8.

Exercise 5.1   Hanning

Verify that equations 1 and 3 imply (5).

Exercise 5.2   Transform of a Tapered Sinusoid

Suppose that $w_0, w_1, \ldots, w_{n-1}$ are a sequence of numbers. In the present context they are interpreted as a data window, but the results of this exercise and the next one do not depend on this interpretation. Their transform is

$$J_w(\omega) = n^{-1} \sum_{t=0}^{n-1} w_t \exp(-i\omega t),$$

and if

$$z_t = w_t \exp(i\lambda t), \qquad t = 0, \ldots, n-1$$

is a tapered sinusoid, verify that the transform of $\{z_t\}$ is

$$J_z(\omega) = n^{-1} \sum_{t=0}^{n-1} z_t \exp(-i\omega t) = n^{-1} \sum_{t=0}^{n-1} w_t \exp\{-i(\omega - \lambda)t\}$$

$$= J_w(\omega - \lambda).$$

Exercise 5.3   Transform of a Product of Series

The following fundamental result is easily verified using the result of the preceding exercise. Suppose that $\{x_t\}$ and $\{y_t\}$ are any two series, with discrete Fourier transforms $\{J_{x,j}\}$ and $\{J_{y,j}\}$, respectively. Let $z_t = x_t y_t$, $t = 0, \ldots, n-1$. If we write x_t as the inverse of its transform

$$x_t = \sum J_{x,k} \exp(i\omega_k t),$$

then z_t is a sum of sinusoids with coefficients $\{J_j\}$, each "tapered" by $\{y_t\}$. Hence deduce that the transform of $\{z_t\}$ is

$$J_{z,j} = \sum_k J_{x,k} J_{y,j-k},$$

the *convolution* of the transforms of $\{x_t\}$ and $\{y_t\}$. In words, *the transform of a product is the convolution of the transforms*. Note that the convolution is defined circularly, since the subscript $j - k$ has to be interpreted modulo n.
   Verify the more general result

$$J_z(\omega) = \sum_k J_{x,k} J_y(\omega - \omega_k)$$

$$= \sum_k J_{y,k} J_x(\omega - \omega_k).$$

Exercise 5.4   Continuation: the Dual Result

The duality between the (complex) discrete Fourier transform and its inverse makes the dual result easy to prove. If

$$z_t = \sum_u x_u y_{t-u}, \qquad t = 0, \ldots, n-1,$$

where now $t-u$ is interpreted *modulo n*, verify that, if ω is a Fourier frequency, then

$$J_z(\omega) = n J_x(\omega) J_y(\omega),$$

or, in words, *the transform of a convolution is the product of the transforms.*

Exercise 5.5   The Split Cosine Bell Data Window

Suppose that $\{u_t\}$ is defined as in (9). Find its transform. (NOTE: As a result of Exercise 5.2, we need to know the transform for all frequencies, not just the Fourier frequencies.) Show that the rate of decay of the transform lies between that of the boxcar and that of the undivided cosine bell (hanning) data window.

5.3   TAPERING THE VARIABLE-STAR DATA

Figure 5.5 is the periodogram (plotted on a logarithmic scale) of the variable-star data, with 25% of the data tapered. Comparison with Figure 5.2 shows that the leakage from the main peaks has been eliminated beyond around the 64th ordinate, that is, more than about 20 ordinates from the higher frequency. In addition, leakage from the subsidiary peaks has been reduced, and hence these are even more clearly defined. There is a penalty, namely, that the peaks themselves are slightly broader or more rounded than they were (see Exercise 5.6), although this change is not visible in the graph. For these data, we conclude that 25% tapering is sufficient to contain the leakage from the main peaks to a usefully narrow band of frequencies, and even reduces some previously unnoticed leakage from the subsidiary peaks. (It should also be pointed out that tapering requires far less thought, programming effort, and computational effort than the model-fitting procedure which led, as one result, to Figure 5.2. However, it gives us no precise information about frequencies, either.)

Figures 5.6 and 5.7 show the results of 50% and 100% tapering, respectively, the latter being the periodogram derived from the hanned transform (6). The progressive reduction of leakage is clear, especially around the

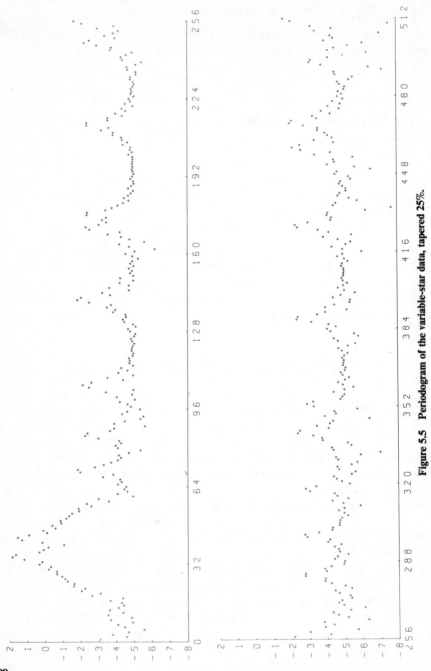

Figure 5.5 Periodogram of the variable-star data, tapered 25%.

88

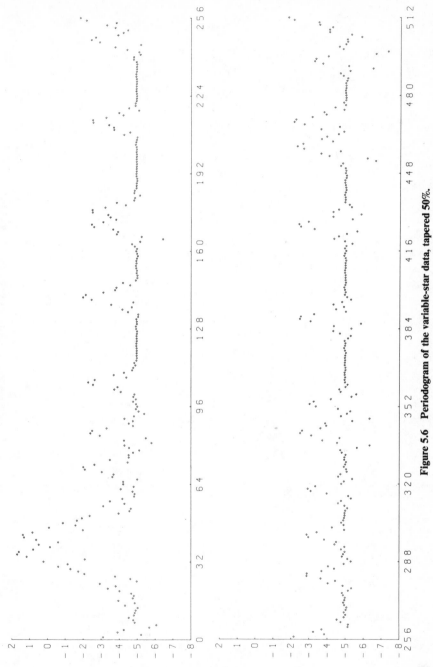

Figure 5.6 Periodogram of the variable-star data, tapered 50%.

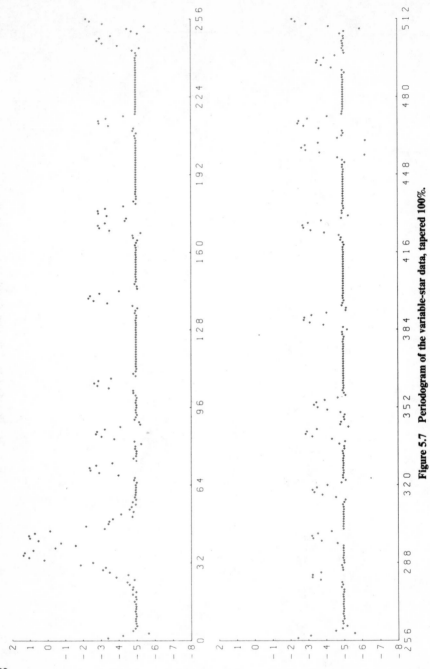

Figure 5.7 Periodogram of the variable-star data, tapered 100%.

main peaks. The accompanying broadening of the peaks may also be seen.

The near constancy of the periodogram away from the peaks is highly unusual. More typically, periodogram values fluctuate randomly in any interval in which there is no peak (see Section 5.8). This suggests that tapering has made the transforms of the periodic components small enough to reveal the presence of yet another component, one with a constant periodogram. The simplest series with a constant periodogram is one in which only a single value is nonzero [see example (ii) of Section 3.4], and usually such a component arises because of an error in a single observation. In this case the phase of the transform is a linear function of frequency (*modulo* 2π), and the slope of the phase plot is the negative of the position of the nonzero term. The magnitude of the nonzero term is n times the magnitude of the transform, or $(n/2)\tilde{R}(\omega)$. The value in this case is therefore around $300 \times (10^{-5})^{1/2} \cong 1$. Since the data are in integers, a likely value for an error is 1, but it would have to be near the middle of the data to keep this value after tapering.

Figure 5.8 is a graph of the phase of the hanned transform, corrected for a slope of $-n/2 = -300$ and reduced *modulo* 2π; that is, the jth ordinate is the residue *modulo* 2π (defined to lie in the interval from $-\pi$ to π) of $\phi_j + 300\omega_j'$, where ϕ_j is the phase of the jth term in the transform. Away from the peaks of the periodogram (and even at the frequencies of some of the peaks), this corrected phase is almost identically 0 and is certainly a linear function of frequency. This confirms that the remaining component is of the "single-error" form (although not finally; we still do not know how its phase behaves at the peaks of the periodogram). Furthermore, the correction has reduced the slope to 0, and therefore the "error" must be in x_{300}, the 301st observation. Finally, since the phase at zero frequency is 0, the "error" is positive. The only other possible value of the phase at zero frequency is π (or, equivalently, $-\pi$), which would correspond to a negative "error" [see example (ii) of Section 3.4].

Had we been less lucky with our choice of a correction of slope, we could tell by how much we missed from the remaining slope. This may be done visually by observing that a slope of $-h$ means that the phase changes downward by $h/2$ multiples of 2π as ω increases from 0 to π. Thus, had we corrected for $t = 295$, say, instead of $t = 300$, the phase plot would have shown a drift from 0 down to $-\pi$, and then two complete cycles from π down to $-\pi$.

Figure 5.9 is the periodogram of the data with $x_{300} = 19$ replaced by $x_{300} - 1 = 18$. (Note that the vertical scaling has been changed to cover the wider range of values.) The troughs now fall to values of around 10^{-8} to 10^{-10}, where previously they were supported by the "floor" at 10^{-5}. The difference between this graph and Figure 5.7 confirms that x_{300} was, in some sense, perturbed.

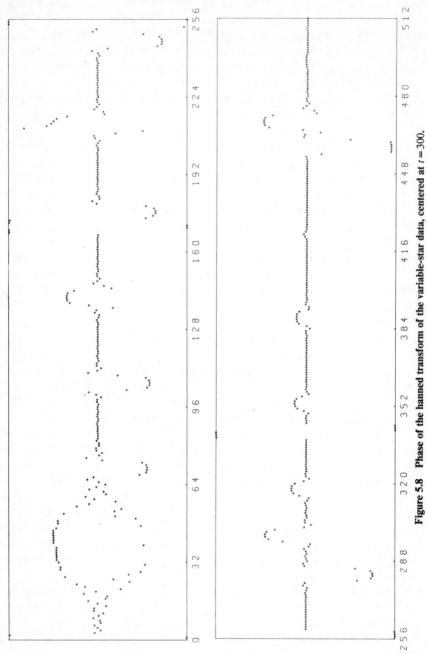

Figure 5.8 Phase of the hanned transform of the variable-star data, centered at $t = 300$.

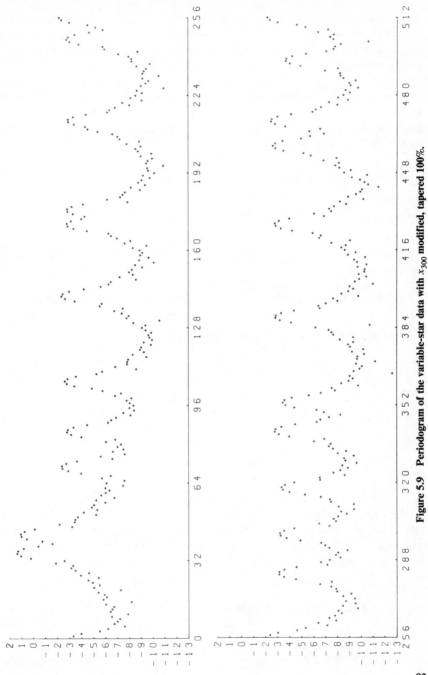

Figure 5.9   Periodogram of the variable-star data with x_{300} modified, tapered 100%.

93

It is interesting that the smallest possible perturbation in a single observation (a change of 1 in the least significant digit) should have such a visible effect on the analysis. This is possible only because the rest of the data have such a strongly periodic character, but a larger perturbation could seriously affect the transform of less highly structured data. This sensitivity to a few "large" errors may be traced to the least squares interpretation of the periodogram developed in Chapter 2 or, equivalently, to the fact that the Fourier transform is a linear function of the data. Any analysis that was less sensitive to errors would necessarily have to be a nonlinear function of the data.

Exercise 5.6   Effect of Tapering on Periodogram Peaks

Consider a sinusoid

$$x_t = \exp(i\lambda t)$$

tapered by w_t, $0 \leqslant t < n$, to give $y_t = w_t x_t$. In the light of Exercise 5.2, all that we need to assume about the data window $\{w_t\}$ is that its transform is nonzero at zero frequency and small elsewhere.

   (i)   Find the ratio of heights of the peaks at $\omega = \lambda$ in the periodograms of $\{x_t\}$ and $\{y_t\}$. In what way should data windows be normalized if we wish to estimate amplitudes without bias?

   (ii)   The curvature of the peak can be measured by the second derivative of the periodogram, evaluated at $\omega = \lambda$. Show that for $\{y_t\}$ this is

$$-\left(\frac{8}{n^2}\right)\left\{ \sum (t - \bar{t})^2 w_t \right\} \sum w_t,$$

where

$$\bar{t} = \frac{\sum t w_t}{\sum w_t}$$

is the average value of t with respect to the weights $\{w_t\}$ [usually, because of symmetry, this is $(n-1)/2$]. Show that this curvature is less than the corresponding curvature for $\{x_t\}$ when the normalization mentioned in (i) has been used (that is, when the peaks are of the same height).

5.4   WOLF'S SUNSPOT NUMBERS

Wolf's sunspot numbers are an index of surface activity of the sun. They are a much-analyzed set of data for which no completely satisfactory explanation exists (see, for instance, Bray and Loughhead, 1964, p. 226, or Newton, 1958, p. 84). It has been shown that solar activity has an impact on many terrestrial series, especially the earth's magnetic field, and its climate. Figure 1.2 shows the annual averages of the sunspot index for the years 1700 to 1960, as tabulated by Waldmeier (1961). The succession of peaks and troughs shows that a definitely periodic phenomenon is at work in the data. For a variety of reasons from scientific curiousity to the need for forecasting future peaks of activity, it is desirable to describe this periodicity as accurately as possible.

The periodogram of these data is shown in Fig. 5.10. The mean of the data was subtracted out, and the data were then tapered 5% at each end and extended by zeros to length 512. The tapering covers 13 years at each end of the data, or a little more than one cycle. This may seem rather small in the light of the improvement made by much heavier tapering in Section

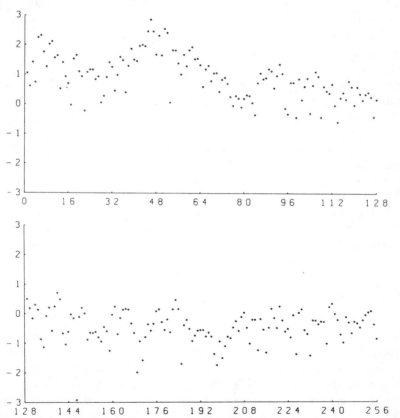

Figure 5.10   Periodogram of the annual sunspot series for frequencies $2\pi j/512, j = 1, \ldots, 257$.

5.3. However, the graph shows that there is in fact no strong peak in the periodogram, and relatively little tapering should be adequate. As usual, the function displayed is the base-10 logarithm of the periodogram [or, rather, of $\tilde{R}(\omega)^2$] for the frequencies ω_j', $0 < \omega_j' \leqslant \pi$.

This periodogram shows a much less strongly defined periodicity than did any periodigrams of the variable-star data in Sections 5.1 and 5.3. The largest ordinate is the 46th, which corresponds to a period of $512/46 \cong 11.13$ years, the commonly quoted "period" of sunspots. However, this ordinate is not dramatically larger than its neighbors, whose magnitude cannot be explained by leakage. There is instead a broad peak with an indeterminate width, extending roughly from the 32nd to the 64th ordinates. Several very low frequencies show almost the same power. Thus it would be misleading to name any single frequency as dominant.

Is there anything in the original data which could warn us that harmonic analysis will not give the clear picture we would like? Recall that a harmonic analysis decomposes the data into sinusoidal terms. The oscillations in these data do in fact show a number of departures from purely sinusoidal behavior. The first is that the sequence of peaks and troughs is not completely regular. For instance, there is no peak between 1787 and 1802, a gap of 15 years, whereas peaks in 1761 and 1769 are separated by only 8 years. The second visible departure involves the amplitude of oscillations. All the troughs occur at around the same level, but the heights of the peaks vary widely. A third feature which may also be seen under close examination is that the individual oscillations are not purely sinusoidal. Typically there is a sharp, well-defined peak and a broad trough, and it often appears that the rise is steeper than the drop. For further description of these features see Newton (1958, pp. 529–554).

In the next two sections we shall describe the general effect on a harmonic analysis of such departures from sinusoidal behavior. In Section 5.7 we examine the extent to which the oscillations may be made more sinusoidal by a suitable transformation.

5.5 NONSINUSOIDAL OSCILLATIONS

The theory of Fourier series shows that any periodic function can be represented as the sum of a (usually infinite) series of sines and cosines. For the discrete Fourier transform described in Chapter 3 there is a corresponding result, given as example (vi) of Section 3.4. It states that, if a series $\{x_t\}$ is periodic with period h, its transform is nonzero only at the frequency $2\pi/h$ and its multiples. (Strictly, for discrete time series, periodicity can be defined only for *integer* periods h, and the result holds only if the series length n is a multiple of the period. However, when the data are obtained by sampling a continuous time phenomenon, a noninteger period has an obvious interpretation and usefulness. If n is not a multiple of the period, a corresponding approximate result holds.) The frequency $2\pi/h$ is the *fundamental* frequency of the oscillation, and its multiples are *harmonics*.

In Section 5.4 we noted some evidence of nonsinusoidal behavior in the sunspot series, Figure 1.1. The first is that the troughs tend to be flatter than the peaks, and the second is that the rises tend to be slightly steeper than the falls. Figure 5.11 shows two periodic functions, each displaying one of these phenomena. The upper curve is $\cos u + \frac{1}{4}\cos 2u$, and the lower is $\cos u - \frac{1}{8}\sin 2u$. Thus each contains just its fundamental and second harmonic.

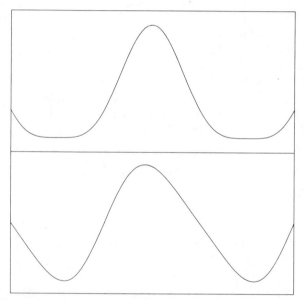

Figure 5.11 **Two periodic functions:** $\cos u + \frac{1}{4}\cos 2u$ **(upper curve) and** $\cos u - \frac{1}{8}\sin 2u$ **(lower curve).**

This suggests that the periodogram of the sunspot data (Figure 5.10) should show some power at the second harmonic of the fundamental frequency. There is indeed some evidence of a second ill-defined peak around the 90th ordinate, but it is very slight. Periodogram smoothing techniques, described in Chapter 7, give us a clearer picture of these broad peaks.

The periodogram provides information only about the *magnitude* of terms in the transform. In the present case, the *direction* of the departure from sinusoidal behavior should be evident in the phase of the transform. If we write the functions in Figure 5.11 as $\cos u + R\cos(2u+\psi)$, the phase ψ has the values 0 and $\pi/2$, respectively. An intermediate phase gives a function showing both types of behavior (see Exercise 5.7). If we change the scale of the u-axis by a factor ω and center the function at $u = u_0$ instead of $u = 0$, it becomes

$$\cos\omega(u - u_0) + R\cos\{2\omega(u - u_0) + \psi\} = \cos(\omega u + \phi_1) + R\cos(2\omega u + \phi_2),$$

say, whence $\psi = \phi_2 - 2\phi_1$, the *relative phase* of the second harmonic.

Since the sunspot periodogram shows a broad peak rather than a single frequency, we computed this relative phase for each frequency in a band covering the peak. The frequencies chosen were $\omega'_j = 2\pi j/512, j = 32,\ldots,64$. If the transform at these frequencies is $J(\omega'_j) = R'_j \exp(i\phi'_j)$, the relative phases are $\psi_j = \phi'_{2j} - 2\phi'_j, j = 32,\ldots,64$.

A natural way to present these numbers graphically is as points on a circle. This corresponds to plotting $\exp(i\psi_j), j = 32,\ldots,64$, in the complex plane. Since we are more interested in frequencies for which the fundamental and second harmonic are strong, it is desirable for the plot to contain also some information about the magnitudes of these terms of the transform. This can be achieved by displacing each point radially by a suitable amount. A convenient quantity is $R'^2_j R'_{2j}$, since

$$R'^2_j R'_{2j} \exp(i\psi_j) = R'^2_j \exp(-2i\phi'_j) R'_{2j} \exp(i\phi_{2j})$$

$$= J(\omega'_{2j}) J(\omega'_j)^{*2}, \tag{10}$$

and hence the real and imaginary parts may be computed easily. Figure 5.12 shows the resulting graph. The cluster of points near the origin corresponds to frequencies in which either the fundamental or the second harmonic is weak. The only points at any distance from the origin are all in or close to the expected quadrant.

Function 10 is a special case of the *third-order periodogram* $I(\omega,\lambda)$, which is proportional to $J(\omega)J(\lambda)J(\omega+\lambda)^*$. In fact, higher-order periodograms may be defined to an arbitrarily high level in a similar way, although they are rarely used. Brillinger and Rosenblatt (1967b) describe a similar analysis of monthly sunspot data. The quantities they compute are smoothed third- and fourth-order periodograms, computed by techniques similar to those described in Chapter 7 for smoothing the conventional (or, in this terminology, second-order) periodogram.

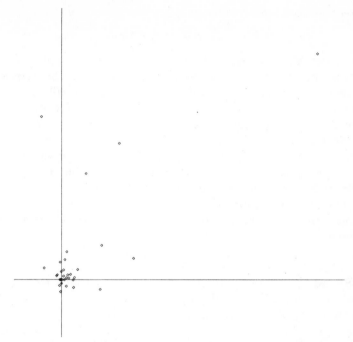

Figure 5.12   Some values of the third-order periodogram of the sunspot series, showing the relative phase of the second harmonic.

Exercise 5.7   *Relative Phase of the Second Harmonic*

Consider the function

$$\cos u + \frac{a}{4} \cos 2u - \frac{b}{8} \sin 2u,$$

where $a, b > 0$ and $a + b = 1$.

   (i)   Show that this function may be written in the form $\cos u + R\cos(2u + \psi)$, where $0 < \psi < \pi/2$.

   (ii)   Show that this function combines the features of the two curves in Figure 5.11. (*Hint*: Evaluate the first derivative at $\pm \pi/2$, and the second derivative at 0 and $\pm \pi$.)

5.6   AMPLITUDE AND PHASE FLUCTUATIONS

The simplest example of a cosine wave with a fluctuating amplitude is the phenomenon of *beats*. If we superimpose two cosine waves with nearly

RICHARD THORPE

equal frequencies $\omega \pm \delta\omega$, the result is

$$\cos(\omega + \delta\omega)t + \cos(\omega - \delta\omega)t = 2\cos\omega t \cos\delta\omega t.$$

This oscillates at the average frequency ω, but the amplitude changes slowly, according to the *modulating function* $2\cos\delta\omega t$. The period of the modulating function is $2\pi/\delta\omega$, which is large if $\delta\omega$ is small. Conversely, if we carry out a Fourier analysis of the modulated cosine wave, we shall find that the apparent frequency ω splits into the two original frequencies $\omega \pm \delta\omega$. In this simple case, the transform in fact is *zero* at frequency ω.

In the case of the sunspot data (Figure 1.2), the modulation is not as simple. In particular, the amplitude seems to vary about a positive value, rather than about 0, as in the case of beats. In addition, it appears that the frequency of the oscillations is not constant, but similarly changes slowly with time. Thus we might expect that the data could be represented approximately as

$$x_t = R_t \cos(\omega_t t + \phi), \qquad (11)$$

where R_t and ω_t are slowly varying sequences (by which terminology we imply no precise mathematical property). If ω_t varies about some typical value λ, this may also be written as

$$x_t = R_t \cos\{\lambda t + \phi + (\omega_t - \omega)t\}$$

$$= R_t \cos(\lambda t + \phi_t), \qquad \text{say}, \qquad (12)$$

in which the frequency is constant and the phase varies. The frequency λ need not be a typical value of ω_t to carry out this rearrangement, but it makes more sense if this is the case, since otherwise the phase ϕ_t would show a systematic drift.

The complex analog of (12) is

$$x_t = R_t \exp\{i(\lambda t + \phi_t)\}$$

$$= z_t \exp(i\lambda t), \qquad (13)$$

say, where $z_t = R_t \exp(i\phi_t)$ is a complex-valued modulating function. Expression 13, a *modulated sinusoid*, is formally identical to the tapered sinusoid of Exercise 5.2, and therefore its transform consists of that of $\{z_t\}$ centered about frequency λ. Specifically,

$$J_x(\omega) = \frac{1}{n}\sum x_t \exp(-i\omega t)$$

$$= \frac{1}{n}\sum z_t \exp\{-i(\omega - \lambda)t\}$$

$$= J_z(\omega - \lambda).$$

Now $\{z_t\}$ varies slowly and smoothly, and hence its transform is large only for low frequencies (see Section 3.5). Therefore the transform of $\{x_t\}$ is large only for frequencies close to λ. However, the single spike that we would see for an unmodulated sinusoid is split or spread into an image of the transform of $\{z_t\}$. Typically the amplitude of the transform of $\{z_t\}$ shows an irregular peak surrounding zero frequency, and in particular rarely vanishes at zero frequency. Thus the transform of $\{x_t\}$ is usually nonzero at frequency λ, but this may not be the largest value.

In the sunspot data we can identify four or five intervals where peaks tend to be large, separated by intervals where they are smaller. The phase variation is harder to see. If we assume that it is changing no more rapidly than the amplitude, we might expect the transform of the (complex) modulating function to be large only for frequencies in the range $\pm\omega_5$. Now $\omega_5 = 2\pi \times 5/261 = 2\pi(5 \times 512/261)/512 \cong 2\pi \times 10/512 = \omega'_{10}$. (Recall that the data were extended to length $n' = 512$.) Thus we might expect the periodogram of the sunspot numbers to show a peak with a width of around 20 ordinates. Figure 5.10 shows a peak of roughly this width, a fact which suggests that the poor definition of that peak may well be due to the amplitude and phase fluctuations. *Complex demodulation*, described in Chapter 6, is a useful tool for the analysis of data containing amplitude or phase fluctuations.

5.7 TRANSFORMATIONS

In Sections 5.5 and 5.6 we described two common ways in which a periodic series may fail to be purely sinusoidal. Both have the effect that the transform of the series has power at frequencies other than the fundamental frequency corresponding to the period of the series. To this extent, there is information about the periodicity of the series in its Fourier transform at frequencies other than the fundamental. In particular, if we look at any one ordinate of the transform, we are not using all of the available information.

A series can sometimes be transformed in such a way that it becomes more closely sinusoidal. This means that a greater proportion of the power associated with the periodicity appears at the fundamental frequency in the transform, and hence it is hoped, the fundamental becomes more pronounced. For example, the upper curve in Figure 5.11 is

$$\cos u + \tfrac{1}{4}\cos 2u = \cos u + \tfrac{1}{4}\left\{2(\cos u)^2 - 1\right\}$$

$$= \tfrac{1}{2}(1 + \cos u)^2 - \tfrac{3}{4}.$$

Thus, if the periodic signal were

$$\tfrac{3}{4} + \cos u + \tfrac{1}{4} \cos 2u,$$

which, like the sunspot data, is nonnegative and has zero as its lowest value, we could transform it into a pure sinusoid by a square root transformation.

As a second example, suppose that a series is $x_t = y_t^2$, where

$$y_t = R \cos(\omega t + \phi) + z_t,$$

and $\{z_t\}$ is a slowly varying series. Then

$$x_t = R^2 \cos(\omega t + \phi)^2 + 2z_t R \cos(\omega t + \phi) + z_t^2$$

$$= \tfrac{1}{2} R^2 \{\cos 2(\omega t + \phi) + 1\} + 2z_t R \cos(\omega t + \phi) + z_t^2, \tag{14}$$

which consists of a term with frequency 2ω, a slowly varying term $\tfrac{1}{2} R^2 + z_t^2$, and an amplitude-modulated term

$$2z_t R \cos(\omega t + \phi).$$

Thus, in this case, the square root transformation would eliminate both the term at 2ω (the second harmonic) and the amplitude modulation. Of course, the original data have to have special structure (14) for this to be the case.

Notice that a transformation of the values of a series can do nothing to remove time asymmetry in a waveform, such as that shown by the lower curve of Figure 5.11. Similarly, variations in the phase of an otherwise periodic signal cannot be changed. These features could be removed only by a transformation of the time axis, which would in general be difficult to find and hard to interpret.

Figure 5.13 shows the square roots of the sunspot numbers. The tendency for the troughs to be flatter than the peaks has been largely removed and possibly partly reversed. The values at the troughs show greater variation than before, although still not as great as that of the heights of the peaks. It also appears that the amplitude modulation has been reduced and a low-frequency component introduced, as in the example above. These observations suggest that the power at the second harmonic should be reduced, and that the main peak should be more clearly defined.

Figure 5.14 shows the periodogram of the square roots, which is in fact disappointingly similar to the original periodogram, Figure 5.10. Close inspection suggests that power at the first harmonic has been reduced slightly, but the main peak is no clearer. The phase fluctuations, which are

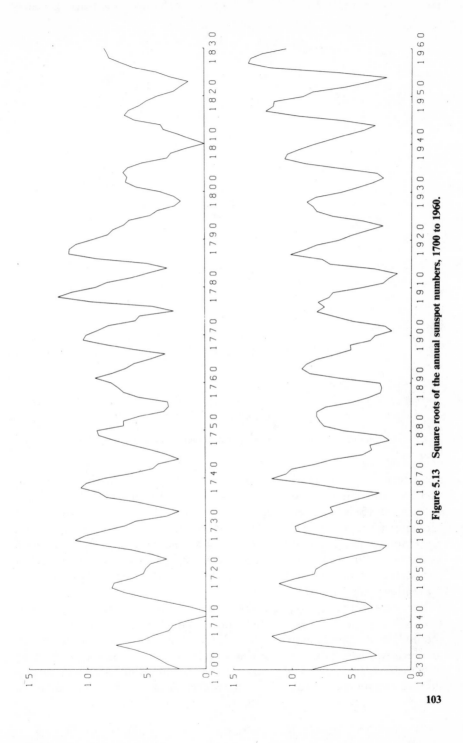

Figure 5.13   Square roots of the annual sunspot numbers, 1700 to 1960.

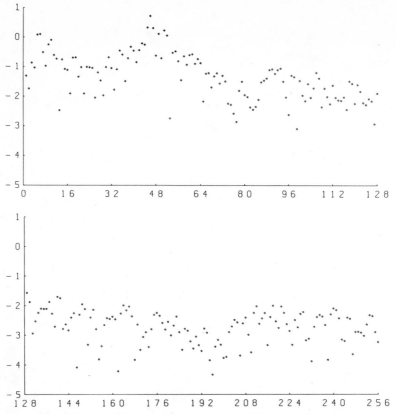

Figure 5.14   Periodogram of the square roots of the sunspot numbers.

not affected by this transformation, and the remaining amplitude fluctuation are sufficiently strong for the peak still to be diffuse rather than sharp.

Figure 5.15 is the relative phase plot for the second harmonic of
fundamentals $\omega_j' = 2\pi j/512, j = 32, \ldots, 64$, computed as described in Section
5.5. Only two points fall at large radii ($j = 45$ and 46), and both of these are
in the second quadrant. This tends to confirm that our transformation has
partly reversed the nature of peaks and troughs, while leaving the asymmetry relatively unchanged.

The most generally useful transformations are the power transformations $y = x^\alpha$, and the logarithmic transformation $y = \ln x$ [which arises as
the limit as α approaches 0 of $(x^\alpha - 1)/\alpha$]. These may be applied only to
nonnegative data (positive data if $\alpha \leqslant 0$). However, the data most in need
of transformation are usually nonnegative, so this is not a serious restric-
tion. In certain special cases other transformations may be suitable. In the

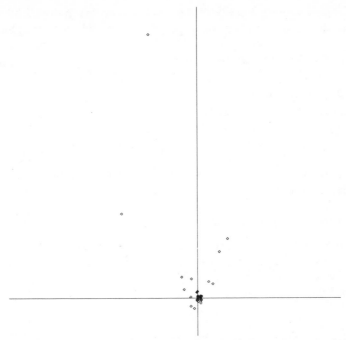

Figure 5.15 Some values of the third-order periodogram of the square roots of the sunspot numbers, showing the relative phase of the second harmonic.

case of the sunspot data, it is doubtful whether another transformation would bring about a substantial improvement. We have seen that the amplitude modulation has not been eliminated, and it appears that a stronger transformation, perhaps cube roots, is needed for this. On the other hand, the natures of the peaks and troughs seem to have been partly reversed, and a stronger transformation could only intensify this undesirable effect.

Exercise 5.8 The Effect of a Transformation

Suppose that $\{y_t: t=0,\ldots,n-1\}$ is obtained from $\{x_t\}$ by a transformation $y_t = f(x_t)$, with inverse $x_t = g(y_t)$. Suppose also that

$$y_t = \mu + R\cos(\omega_j t + \phi),$$

where ω_j is the jth Fourier frequency.

   (i) Suppose in addition that j divides n. Show that the transform of $\{x_t\}$ is nonzero only at multiples of ω_j.

(ii)   Suppose more generally that the greatest common divisor of j and n is l. Show that the transform of $\{x_t\}$ is nonzero only at multiples of ω_l. (*Hint*: Find the period of $\{y_t\}$, and hence of $\{x_t\}$, and note that the period of a discrete time sequence must be an integer.)

Exercise 5.9   Continuation

Suppose that $\{x_t\}$ and $\{y_t\}$ are related as in Exercise 5.8, but that

$$y_t = \mu + R_1 \cos(\omega_j t + \phi_1) + R_2 \cos(\omega_k t + \phi_2).$$

Show that the transform of $\{x_t\}$ is nonzero only at multiples of ω_l, where l is the greatest common divisor of j, k, and n.

Exercise 5.10   An Approximation

Suppose that $\{x_t\}$ and $\{y_t\}$ are as in Exercise 5.9, and that R_1 and R_2 are small compared with μ. Use a Taylor series expansion for $g(x)$ about μ to show that in the transform of $\{x_t\}$ there are first-order terms at ω_j and ω_k, and second-order terms at $2\omega_j$, $2\omega_k$, and $|\omega_j \pm \omega_k|$.

5.8   THE PERIODOGRAM OF A NOISE SERIES

One feature common to most of the periodograms shown in this chapter is the roughness of the graphs in any interval in which there is no peak. The exceptions were the periodograms of the heavily tapered variable-star data, which showed an unusual constancy (Figures 5.6 and 5.7). Even in this case, however, the characteristic roughness reappeared when the exceptional value was corrected (Figure 5.9). This behavior is typical for periodograms of empirical data and is usually caused by the presence of a random component in the data.

The simplest such component is a series of independent random numbers. Figure 5.16 shows 128 pseudorandom numbers drawn from the normal distribution. Their periodogram, shown in Figure 5.17, manifests the typical roughness, but otherwise has no interesting structure or overall shape. By analogy with the spectrum of white light, which is also flat and featureless, a series of independent errors is also known as *white noise*.

Figure 5.18 is the histogram of this periodogram, with an exponential curve superimposed. The similarity is no coincidence. It may be shown that, when the data are independent random numbers drawn from a normal distribution, the periodogram ordinates at the Fourier frequencies (other than 0 and π) are independently exponentially distributed (see

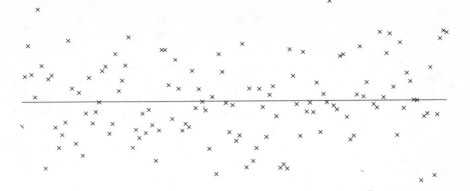

Figure 5.16   A series of 128 pseudorandom deviates from the standard normal distribution.

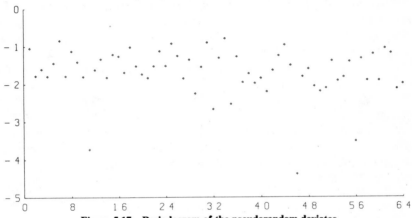

Figure 5.17   Periodogram of the pseudorandom deviates.

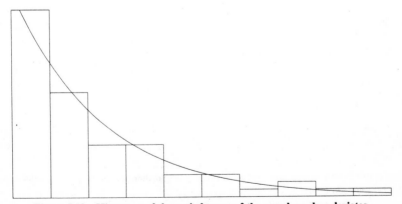

Figure 5.18   Histogram of the periodogram of the pseudorandom deviates.

Exercise 5.11). It is this independence that makes the graph of the periodogram rough, since it means that there is no tendency for adjacent ordinates to have similar values.

When the periodogram is calculated at frequencies other than the Fourier frequencies (for instance, when it is calculated using the fast Fourier transform on a series extended by zeros), each ordinate is still approximately exponentially distributed, but neighboring ordinates are by no means independent (see Exercise 5.12). Tapering a series adds a further complication, since then the distribution of the periodogram ordinates is only approximately exponential even at the Fourier frequencies (see Exercise 5.13).

Figure 5.19 is the histogram of the last 64 ordinates in the periodogram of the sunspots (Figure 5.10), again with an exponential curve superimposed. Despite the fact that the data were both tapered and extended by zeros, no gross departure from the exponential distribution is evident. The number of ordinates used, 64, is too small to detect slight departures. No more ordinates could be used, however, since the sunspot periodogram begins to show a gradual increase as ω moves back toward $\pi/2$ ($j = 128$). Inclusion of these ordinates would introduce an exponential distribution with a different scale parameter.

The exponential distribution of periodogram ordinates has been derived here on the assumption of independent, normally distributed errors (or *noise*). The key point, however, is that the cosine and sine sums have normal distributions. If we assume only that the errors are independent and have finite variance, the central limit theorem (see, for instance, Feller, 1968, pp. 244, 254) assures us that these sums are approximately normally

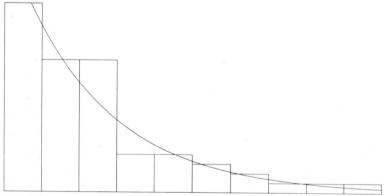

Figure 5.19   Histogram of the last 64 ordinates of the periodogram of the annual sunspot numbers (shown in Figure 5.10).

distributed for large n. It follows that periodogram ordinates are still approximately exponentially distributed.

Exercise 5.11   Periodogram of a Noise Series

Suppose that $x_0, \ldots, x_{n-1}$ are independent and identically distributed with the standard normal distribution. If ω is a Fourier frequency, $\omega \neq 0$ or π, show that both

$$\sum_{t=0}^{n-1} x_t \cos \omega t \quad \text{and} \quad \sum_{t=0}^{n-1} x_t \sin \omega t$$

have zero expectation, that their variances are both $n/2$, and that their covariance is zero. It follows that the periodogram has a χ^2-distribution with 2 degrees of freedom, which is the exponential distribution. (Note that, when the periodogram is computed at the Fourier frequencies, it makes no difference whether the mean is subtracted, apart from numerical errors.)

Exercise 5.12   Periodogram of an Extended Noise Series

Suppose that $\{x_t\}$ is as in Exercise 5.11, but that ω is not a Fourier frequency. Show that the values found in Exercise 5.11 for the moments of the cosine and sine sums are still approximately valid, and deduce that the periodogram is approximately exponentially distributed. (See also Exercise 2.7 to 2.9.)

Exercise 5.13   Periodogram of a Tapered Noise Series

Suppose that $\{x_t\}$ is as in Exercise 5.11, and consider the cosine and sine sums of the series tapered by $\{w_t\}$,

$$\sum_{t=0}^{n-1} w_t x_t \cos \omega t, \quad \sum_{t=0}^{n-1} w_t x_t \sin \omega t.$$

Suppose further that $w_t = w\{(t + \frac{1}{2})/n\}$ for some smooth function $w(x)$, defined for $0 \leqslant x \leqslant 1$. Find the variances and covariance of these sums, and show that they may be approximated by integrals involving $w(x)^2$. Use the fact that

$$\int_0^1 w(x)^2 \cos Ax \, dx, \quad \int_0^1 w(x)^2 \sin Ax \, dx$$

converge to 0 as $A \to \infty$ to deduce as in Exercise 5.12 that the periodogram is approximately exponentially distributed.

5.9 FISHER'S TEST FOR PERIODICITY

The distributional results of the preceding section show that it may be misleading to look for peaks in a periodogram. Since the ordinates at the Fourier frequencies are approximately independent, they are bound to fluctuate and show many small peaks and troughs. Furthermore, since the distribution is exponential, the largest ordinate, even for a noise series, tends to be large compared with its neighbors (see Exercise 5.14) and may appear to indicate a reasonably strong periodicity.

Fisher (1929) has proposed a test of the significance of the largest peak in a periodogram and gives a table of critical values for various series lengths. The test has been extended by several authors to include testing the second largest peak and others. Shimshoni (1971) gives tables for various situations. The original test statistic is the largest of the periodogram ordinates at Fourier frequencies divided by the sum of these ordinates. The critical values tabulated are percentiles of the distribution

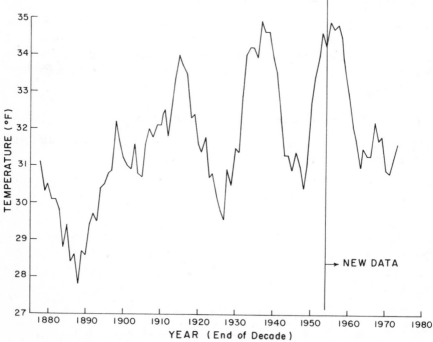

Figure 5.20   Ten-year running means of mean January temperatures at Central Park Observatory, New York City, for 1869 to 1973 (from Spar and Mayer, 1973).

of this statistic under the null hypothesis that the series consists of independent errors. (For an exact theory, it is assumed that the errors are normally distributed, but, as in Section 5.8, if this assumption is not valid the theory provides an approximation.)

This null hypothesis is often inappropriate, in that it may be clear from the data that, even if no periodicity is present, the data do not consist of independent errors. Whittle (1952) describes a modified procedure that is appropriate for a more general null hypothesis.

As an example of the application of the original test, consider the data shown in Figure 5.20, which are 10-year running means of January temperatures for 1869 to 1973 at Central Park Observatory, New York City (from Spar and Mayer, 1973). There appears to be a strong 20-year cycle, especially in the more recent observations. Figure 5.21 shows the periodogram of the *original*, unsmoothed, January temperatures. (We emphasize that this is *not* the periodogram of the data in Figure 5.20, which, being a relatively smooth series, would have more power at low frequencies. Notice also that the periodogram is plotted on a linear rather than a logarithmic scale. The periodogram ordinates at the Fourier frequencies have been interpolated linearly, thus providing a rough ap-

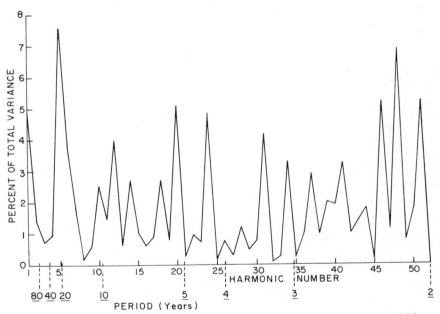

Figure 5.21   Periodogram of unsmoothed mean January temperatures (from Spar and Mayer, 1973).

proximation to the graph of the periodogram as a function of a continuous variable ω.) The units of the vertical axis are "percent of total variance." By the sum of squares identities described in Section 3.3, the total variance is (proportional to) the sum of the periodogram ordinates at the nonzero Fourier frequencies, and thus the scale is equivalent to "percent of sum of ordinates." We see therefore that in this case Fisher's test statistic has the value 0.075. The closest entries in Shimshoni's table are for 50 periodogram ordinates, for which the 95% point of the distribution of the statistic under the null hypothesis is 0.131. In fact the value 0.075 is well in the middle of the distribution (see Exercises 5.14). Thus the peak corresponding to a 20-year cycle is far short of statistical significance. It follows that these data alone provide no strong evidence for the presence of a 20-year cycle.

Exercise 5.14   The Largest Periodogram Ordinate

Suppose that $I_1, \ldots, I_m$ are independent and exponentially distributed, that is,

$$\mathrm{pr}(I_j \leqslant x) = 1 - e^{-x}.$$

Let $X_m = \max\{I_j; j = 1, \ldots, m\}$.
  (i)   Show that

$$\mathrm{pr}(X_m \leqslant x) = (1 - e^{-x})^m,$$

and hence that

$$\mathrm{pr}(X_m \leqslant x + \ln m) \cong \exp(-e^{-x})$$

for large m. This implies that X_m is probably close to $\ln m$.
  (ii)   Show that for large m

$$Y_m = I_1 + \cdots + I_m \cong m,$$

in the sense that the distribution of Y_m/m becomes concentrated arbitrarily close to the value 1.
  (iii)   Deduce that, if $G_m = X_m/Y_m$ is Fisher's statistic, then

$$\mathrm{pr}(mG_m \leqslant x + \ln m) \cong \exp(-e^{-x}).$$

For the temperature data of Figure 5.20, $m = 104/2 = 52$. The observed value of 0.075 corresponds to

$$x = m \times 0.075 - \ln m$$
$$= 52 \times 0.075 - \ln 52$$
$$= 3.9 - \ln 52,$$

and hence

$$\exp(-e^{-x}) = \exp\{-e^{-(3.9 - \ln 52)}\}$$
$$= \exp(-52e^{-3.9})$$
$$= \exp(-1.05)$$
$$= 0.35.$$

Thus the value 0.075 would be exceeded by the largest ordinate approximately 65% of the time, even for a pure noise series.

APPENDIX

The following program was used to compute the periodograms displayed in this chapter. Subprogram DETRND is used to subtract the series mean or the least squares regression line from the data, according to the value of the parameter K. For greater generality it could be replaced by a general polynomial regression routine. In some instances it may be preferable to use a different form of detrending, for example, by fitting and subtracting a spline function. Subprogram TAPER implements the split cosine bell data window described in Section 5.2. Subprogram DATOUT is used to output the transform and the periodogram in a format suitable for future reading by DATIN.

The periodogram is computed as the squared amplitude $\tilde{R}(\omega)^2$, rather than the true periodogram

$$I(\omega) = \frac{n}{8\pi} \tilde{R}(\omega)^2.$$

NOTE: The program also uses subprogram DATIN (presented in the Appendix to Chapter 2).

```
C
C     THIS PROGRAM COMPUTES THE DISCRETE FOURIER TRANSFORM
C     OF A SERIES,  AND THE PERIODOGRAM.   THE PROGRAM IS
C     CONTROLLED BY THE FOLLOWING VARIABLES,  WHICH ARE
C     INPUT TO THE PROGRAM  -
C
C     P     THE TOTAL PROPORTION OF THE DATA TO BE TAPERED
C     K     THE DEGREE OF THE POLYNOMIAL TO BE REMOVED (0 OR 1)
C
      DIMENSION X(1024),Y(1024)
      DO 10 I=1,1024
      X(I)=0
 10   Y(I)=0
      CALL DATIN (X,N,START,STEP,7)
      READ(5,1) K
 1    FORMAT(10I5)
      K=MAXO(0,MINO(1,K))
      WRITE(6,2) K
 2    FORMAT(≠0THE LEAST SQUARES POLYNOMIAL OF DEGREE≠,I5,
     +          ≠ WAS REMOVED FROM THE DATA≠)
      CALL DETRND (X,N,K)
      READ(5,3) P
 3    FORMAT(F10.5)
      WRITE(6,4) P
 4    FORMAT(≠0TOTAL PROPORTION OF DATA TAPERED IS≠,F6.3)
      CALL TAPER (X,N,P)
      NP2=1
 20   NP2=NP2*2
      IF (NP2 .LT. N) GO TO 20
      INV=0
      CALL FT01A (X,Y,NP2,INV)
      IF (INV .EQ. 0) GO TO 30
      WRITE(6,5)
 5    FORMAT(≠0    TROUBLE IN FT01A  --  NP2 OUT OF RANGE≠)
      STOP
 30   CONTINUE
      NPGM=NP2/2+1
      CALL DATOUT (X,NPGM,0.0,1.0,8)
      CALL DATOUT (Y,NPGM,0.0,1.0,8)
      CON=(2.0*FLOAT(NP2)/FLOAT(N))**2
      DO 40 I=1,NPGM
 40   X(I)=(X(I)**2+Y(I)**2)*CON
      CALL DATOUT (X,NPGM,0.0,1.0,9)
      STOP
      END
```

```
      SUBROUTINE DETRND (X,N,K)
C
C     THIS SUBROUTINE COMPUTES AND SUBTRACTS FROM THE
C     SERIES  X  EITHER THE SERIES MEAN OR THE LEAST SQUARES
C     STRAIGHT LINE.   NOTE THAT THESE ARE THE LEAST SQUARES
C     POLYNOMIALS OF DEGREE  0  AND  1,  RESPECTIVELY.
C     PARAMETERS ARE
C
C     NAME    TYPE                            VALUE
C                                 ON ENTRY                ON RETURN
C
C      X      REAL ARRAY THE SERIES           DETRENDED SERIES
C
C      N       INTEGER    SERIES LENGTH       UNCHANGED
C
C      K       INTEGER    DEGREE OF POLYNOMIAL    UNCHANGED
C                         TO BE FITTED
C
       DIMENSION X(N)
       SUMX=0.0
       DO 20 I=1,N
   20  SUMX=SUMX+X(I)
       XBAR=SUMX/FLOAT(N)
       DO 30 I=1,N
   30  X(I)=X(I)-XBAR
       IF(K .LE. 0) RETURN
       TBAR=FLOAT(N+1)/2.0
       SUMTT=FLOAT(N*(N*N-1))/12.0
       SUMTX=0
       DO 40 I=1,N
   40  SUMTX=SUMTX+X(I)*(FLOAT(I)-TBAR)
       BETA=SUMTX/SUMTT
       DO 50 I=1,N
   50  X(I)=X(I)-BETA*(FLOAT(I)-TBAR)
       RETURN
       END
```

```
      SUBROUTINE TAPER (X,N,P)
C
C   THIS SUBROUTINE APPLIES SPLIT-COSINE-BELL TAPERING TO
C   THE TIME SERIES X.  THE ARGUMENT  P  IS THE TOTAL
C   PROPORTION OF THE SERIES WHICH IS TAPERED.
C   PARAMETERS ARE
C
C   NAME    TYPE                        VALUE
C                           ON ENTRY              ON RETURN
C
C   X     REAL ARRAY THE TIME SERIES          THE SERIES AFTER
C                                             TAPERING
C
C   N     INTEGER    THE SERIES LENGTH        UNCHANGED
C
C   P     REAL       THE TOTAL PROPORTION     UNCHANGED
C                    OF THE SERIES TO BE
C                    TAPERED
C
      DATA PI /3.141593/
      DIMENSION X(N)
      IF ((P .LE. 0.0) .OR. (P .GT. 1.0)) RETURN
      M=INT(P*FLOAT(N)+0.5)/2
      DO 10 I=1,M
      WEIGHT=0.5-0.5*COS(PI*(FLOAT(I)-0.5)/FLOAT(M))
      X(I)=X(I)*WEIGHT
      X(N+1-I)=X(N+1-I)*WEIGHT
   10 CONTINUE
      RETURN
      END
```

```
      SUBROUTINE DATOUT (X,N,START,STEP,M)
C
C   THIS SUBROUTINE OUTPUTS THE SERIES  X   IN THE FORMAT
C   EXPECTED BY SUBROUTINE DATIN.   THE HEADER CARD IS
C   BLANK EXCEPT FOR SEQUENCING.   PARAMETERS ARE -
C
C   NAME    TYPE              VALUE  (NONE ARE CHANGED)
C
C   X       REAL ARRAY        THE SERIES
C
C   N       INTEGER           SERIES LENGTH
C
C   START   REAL              TIME ORIGIN
C
C   STEP    REAL              TIME INCREMENT
C
C   M       INTEGER           LOGICAL UNIT NUMBER
C
      DIMENSION X(N),Z(4)
      DATA Z /4*0.0/
      K=1
      WRITE(M,1) K
 1    FORMAT(75X,I5)
      K=2
      WRITE(M,2) N,K
 2    FORMAT(I5,70X,I5)
      K=3
      WRITE(M,3) K
 3    FORMAT(≠ (5E15.8)≠,66X,I5)
      K=4
      WRITE(M,4) START,STEP,K
 4    FORMAT(2F10.5,55X,I5)
      L=5
      IHI=0
 10   ILO=IHI+1
      IHI=IHI+L
      K=K+1
      IF (IHI .GT. N) GO TO 20
      WRITE(M,5) (X(I),I=ILO,IHI),K
 5    FORMAT(5E15.8,I5)
      GO TO 10
 20   IF (ILO .GT. N) RETURN
      LIM=IHI-N
      WRITE(M,5) (X(I),I=ILO,N),(Z(I),I=1,LIM),K
      RETURN
      END
```

6

COMPLEX DEMODULATION

Harmonic analysis of the sunspot data in Chapter 5 showed that not all "periodic" phenomena have simple representations in terms of cosine functions, even when we have done all that we can to improve the analysis. *Complex demodulation* is a more flexible approach to such data (Bingham, Godfrey, and Tukey, 1967). It may be used to describe features of data that would be missed by harmonic analysis, and also to verify in some cases that no such features exist. The price of this flexibility is a loss of precision in describing pure frequencies, for which harmonic analysis is most exact.

6.1 MOTIVATION

Suppose that a set of data contains a perturbed periodic component

$$x_t = R_t \cos(\lambda t + \phi_t). \tag{1}$$

Here $\{R_t\}$ is a slowly changing amplitude, and $\{\phi_t\}$ is a slowly changing phase. It was shown in Section 5.4 that the oscillations in the sunspot data could reasonably be regarded as having the structure (1). The aim of complex demodulation is to extract approximations to the series $\{R_t\}$ and $\{\phi_t\}$. It may be regarded as a *local* version of harmonic analysis; it is analogous to harmonic analysis in that it seeks to describe the amplitude and phase of an oscillation, but it is local in that the amplitude and phase are determined only by the data in the neighborhood of t, rather than by the whole series.

Consider first the complex analog of (1),

$$x_t = R_t \exp\{ i(\lambda t + \phi_t) \}.$$

The extraction of $\{R_t\}$ and $\{\phi_t\}$ is trivial if λ is known, as we assume it is, for

$$y_t = x_t \exp(-i\lambda t)$$
$$= R_t \exp(i\phi_t),$$

and thus

$$R_t = |y_t|, \quad \text{and} \quad \exp(i\phi_t) = \frac{y_t}{|y_t|}.$$

We say that $\{y_t\}$ is obtained from $\{x_t\}$ by *complex demodulation*. Now the real form (1) may be written as

$$x_t = \tfrac{1}{2} R_t \big[\exp\{ i(\lambda t + \phi_t) \} + \exp\{ -i(\lambda t + \phi_t) \} \big] \tag{2}$$

and is thus the sum of two complex terms. If we again let

$$y_t = x_t \exp(-i\lambda t),$$

we find

$$y_t = \tfrac{1}{2} R_t \exp(i\phi_t) + \tfrac{1}{2} R_t \exp\{ -i(2\lambda t + \phi_t) \}.$$

The first term is the one we want, since as before it is trivial to extract the sequences $\{R_t\}$ and $\{\phi_t\}$. The second term, which is a perturbed complex sinusoid with frequency -2λ, has to be removed before we can proceed.

In general the data being analyzed do not consist solely of a perturbed sinusoid. For instance, it was shown in Section 5.4 that the sunspot data contain some low-frequency terms, and in Section 5.5 that they contain some terms which may be identified as second harmonics. Also it is clear from the graph of the data, Figure 5.10, that there is a noise component. Thus, more generally, we would have

$$x_t = R_t \cos(\lambda t + \phi_t) + z_t, \tag{3}$$

where $\{z_t\}$ are the additional terms, and hence

$$y_t = x_t \exp(-i\lambda t)$$
$$= \tfrac{1}{2} R_t \exp(i\phi_t) + \tfrac{1}{2} R_t \exp\{ -i(2\lambda t + \phi_t) \} + z_t \exp(-i\lambda t). \tag{4}$$

The basic problem in complex demodulation is to separate the first term in (4) from the others. The feature which makes this possible is that, since both $\{R_t\}$ and $\{\phi_t\}$ are assumed to be smooth, the first term likewise is smooth. The second term oscillates at a frequency around -2λ. Whatever frequencies are present in the final term are shifted by $-\lambda$ by the complex demodulation (see Exercise 6.1). Now $\{z_t\}$ may be assumed to have no component at frequency λ, since any such component cannot be distinguished from $R_t\cos(\lambda t + \phi_t)$. Thus $\{z_t\exp(-i\lambda t)\}$ contains no component around zero frequency and hence is not smooth (see Section 3.4). The problem is to extract the smooth component of $\{y_t\}$. This is usually accomplished by *linear filtering*, which is described briefly in the next section.

Exercise 6.1   The Transform of a Demodulated Series

For any series $\{x_t\}$, the demodulated series

$$y_t = x_t\exp(-i\lambda t), \qquad t = 0,\dots,n-1$$

is formally the same as the tapered sinusoid of Exercise 5.2. Verify that the transform of $\{y_t\}$ consists of the transform of $\{x_t\}$, centered at λ, that is,

$$J_y(\omega) = n^{-1}\sum y_t\exp(-i\omega t)$$

$$= n^{-1}\sum x_t\exp\{-i(\omega+\lambda)t\}$$

$$= J_x(\omega+\lambda).$$

In this sense, complex demodulation just shifts all the frequencies by $-\lambda$.

6.2   SMOOTHING; LINEAR FILTERING

We have just encountered a problem that arises in many contexts: given data that consist of a smooth function, the *signal*, plus disturbance or *noise*, how can we separate the two components? The question is essentially one of smoothing the data. This is almost always done by the application of a *linear filter*. Since these filters have been discussed extensively elsewhere (see, for instance, Otnes and Enochson, 1972, Chapter 3), we give only a brief account here.

Suppose that the data $\{y_t\}$ can be written as

$$y_t = a_t + e_t,$$

where $\{a_t\}$ is smooth and $\{e_t\}$ represents errors or disturbances. Since $\{a_t\}$ is smooth, a_{t+1} and a_{t-1} will be approximately the same as a_t. Thus, if we average y_{t-1}, y_t, and y_{t+1}, the result will be approximately a_t plus the average of e_{t-1}, e_t, and e_{t+1}. However, these errors will tend to cancel out, so that the average error will tend to be smaller than the individual errors. If we repeat this averaging for each value of t, we obtain a new set of data, say $\{z_t\}$, consisting approximately of $\{a_t\}$ plus errors that tend to be smaller than before. Hence we have gone some way toward extracting the smooth series $\{a_t\}$, except for one point at each end of the data, for which the average cannot be calculated without some modification.

The clearest way to describe the effect of this procedure, which is known as *simple moving averaging*, is through a frequency approach. Suppose first that the data $\{y_t\}$ are exactly sinusoidal,

$$y_t = R\cos(\omega t + \phi).$$

Then

$$z_t = \tfrac{1}{3}(y_{t-1} + y_t + y_{t+1})$$

$$= \tfrac{1}{3}R\left\{\cos(\omega t - \omega + \phi) + \cos(\omega t + \phi) + \cos(\omega t + \omega + \phi)\right\},$$

which we can evaluate most easily as the real part of

$$\tfrac{1}{3}R\left[\exp\{i(\omega t - \omega + \phi)\} + \exp\{i(\omega t + \phi)\} + \exp\{i(\omega t + \omega + \phi)\}\right]$$

$$= \tfrac{1}{3}R\exp\{i(\omega t + \phi)\}\{\exp(-i\omega) + 1 + \exp(i\omega)\}$$

$$= \tfrac{1}{3}R\exp\{i(\omega t + \phi)\}(1 + 2\cos\omega)$$

The real part is thus $\tfrac{1}{3}R\cos(\omega t + \phi)(1 + 2\cos\omega)$. Thus the *output* of this procedure, $\{z_t\}$, is obtained from the *input*, $\{y_t\}$, by multiplying by $\tfrac{1}{3}(1 + 2\cos\omega)$. A graph of this as a function of ω is given in Figure 6.1.

Because in the procedure that we have used the output is a linear function of the input, we can also say what will happen when the input is the sum of a number of cosine terms. The output then will contain cosine terms with the same frequencies, but with amplitudes changed by the factor $\tfrac{1}{3}(1 + 2\cos\omega)$. Thus frequencies near zero pass through relatively undiminished, whereas a term with $\omega = 2\pi/3$ will be removed completely.

However, we saw in Chapter 3 that any set of data can be written as a sum of cosine terms, and hence we can describe the action of this procedure on such arbitrary data in these frequency terms. This provides an alternative description of the effect of simple averaging in the original problem, where y_t is smooth "signal" plus error, or "noise," for we saw in

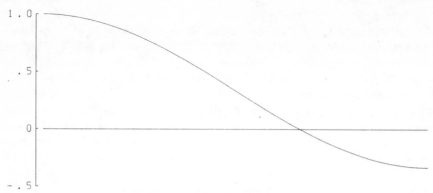

Figure 6.1   The function $(1+2\cos\omega)/3$.

Chapter 3 that for a function to be smooth its transform must be con-centrated at low frequencies, whereas at least for *white* noise (see Section 5.8) the magnitude of the transform is relatively constant. Thus the averaging procedure we have used here will pass most of the signal, but will cut down the power of the noise, at least over certain frequency bands.

A general linear filter consists of a set of *weights* $\{g_r, g_{r+1}, \ldots, g_s\}$, such that, if the input to the filter is $\{y_t\}$, the output is

$$z_t = \sum_{u=r}^{s} g_u y_{t-u}. \tag{5}$$

The three-term simple moving average above has $r = -1$, $s = 1$, and $g_u = \frac{1}{3}$, $u = -1, 0, 1$. Note that, if the input series is available for $t = 0, 1, \ldots, n-1$, the output may be computed only for $t = s, \ldots, n-1+r$. When the input series is the sinusoid $R\cos(\omega t + \phi)$, the output, for t in the latter range, is the real part of

$$\sum_{u=r}^{s} g_u R \exp\{i(\omega t - \omega u + \phi)\} = R \exp\{i(\omega t + \phi)\} \sum_{u=r}^{s} g_u \exp(-i\omega u).$$

The second factor,

$$G(\omega) = \sum_{u=r}^{s} g_u \exp(-i\omega u),$$

is called the *transfer function* of the filter, since it describes the way in which a sinusoid with frequency ω is transferred from the input to the output. Its squared magnitude $|G(\omega)|^2$ is called the *power transfer function*.

In the case of a *symmetric* filter, one for which $g_{-u} = g_u$, the transfer function will be real, and the output is $R\cos(\omega t + \phi)G(\omega)$. Notice that in this case we may write

$$G(\omega) = \sum g_u \cos \omega u$$

$$= g_0 + 2 \sum_{u > 0} g_u \cos \omega u,$$

and since $\cos \omega u$ is a polynomial in $\cos \omega$, $G(\omega)$ may be expanded as a polynomial in $\cos \omega$. More generally, $G(\omega)$ may be complex, say $G(\omega) = \Gamma(\omega)\exp\{i\gamma(\omega)\}$. The output is then the real part of

$$R\exp\{i(\omega t + \phi)\}\Gamma(\omega)\exp\{i\gamma(\omega)\},$$

which is

$$R\Gamma(\omega)\cos\{\omega t + \phi + \gamma(\omega)\}.$$

In this case both an amplitude change of $\Gamma(\omega)$ and a phase change of $\gamma(\omega)$ occur. It should be noticed that when we are operating on real-valued data we shall usually use real-valued coefficients; this means that $G(-\omega) = G(\omega)^*$. Thus $\Gamma(-\omega) = \Gamma(\omega)$, and $\gamma(-\omega) = -\gamma(\omega)$. Hence, as might be expected, we need discuss only the behavior of transfer functions for positive frequencies.

Relation 5 defining the output of the filter in terms of its input is an example of *convolution*. If the values at the ends of the output are computed by assuming that the input is part of a periodic sequence of period n, we have the *circular* convolution operation defined in Exercise 5.4. It follows from the results of that exercise that if ω is a Fourier frequency

$$J_z(\omega) = nJ_y(\omega)J_g(\omega)$$

$$= G(\omega)J_y(\omega), \qquad (6)$$

since the transform of the weights is just $J_g(\omega) = n^{-1}G(\omega)$. In general, some other rule is used to fill in the end values. Two simple rules are to assume that unavailable values are equal to the corresponding end value, and to assume that the series has even symmetry about each end value. It may be shown that for any reasonable rule relation 6 is approximately true (see Exercise 6.3).

Since the transfer function is equivalent to the transform of the weights, it may be inverted to give the weights. Thus a general linear filter may be defined in terms of either the weights $\{g_u\}$ or the transfer function $G(\omega)$. The weights are also known as the *impulse response function*, since if the input series is an *impulse* (that is, contains just a single nonzero value), the output consists just of the weights (see Exercise 6.4). The filters we have described are known as *finite impulse response* (FIR) filters. We shall not use filters outside this class. The other main class consists of *recursive filters*, which do not have finite impulse response. The transfer function of a filter is also known as its *frequency response function*, since in a dual way it describes the output when the input contains just a single frequency.

The simplest filters are those which *shift* the series; that is, if the input is $\{x_t\}$, the output is $\{y_t\}$, where $y_t = x_{t+h}$, for some h. Since $x_0 = y_h$, we may think of this as a change of origin to $t = h$. The transfer function is $\exp(ih\omega)$, whence $\Gamma(\omega) = 1$, $\gamma(\omega) = h\omega$. Whenever a transfer function has a linear phase, we may regard it in this sense as including a shift, even if $\Gamma(\omega)$ is not constant.

A convenient way to construct filters is by repeated application. If we have two filters with transfer functions $G_1(\omega)$ and $G_2(\omega)$, the result of applying one to the output of the other is a filter with transfer function $G_1(\omega)G_2(\omega)$, for if the input is $\exp(i\omega t)$, then the output from the first, which is the input to the second, is $G_1(\omega)\exp i\omega t$. Hence the final output is $G_1(\omega)G_2(\omega)\exp(i\omega t)$. It does not matter in which order the filters are applied, except for effects at the ends of the data. The transfer functions $G_1(\omega)G_2(\omega)$ and $G_2(\omega)G_1(\omega)$ are identical, and by our previous argument the transfer function determines the filter.

Exercise 6.2   Circular Filters

Suppose that a linear filter with weights $\{g_u : r \leqslant u \leqslant s\}$ is defined as a circular convolution. Show that (6) holds approximately even if ω is not a Fourier frequency. [Hint: Write out (6) explicitly for the end terms, and note that only a finite number of terms are affected.]

Exercise 6.3   (Continuation) Noncircular Filters

A possible definition of a *reasonable* rule for extending an input series in order to compute the output of a noncircular filter would be that any substitute value falls in the *range* of the data. In other words, if $\hat{y}$ is any value substituted for y_t, $t < 0$ or $t \geqslant n$, then

$$\min\{y_t : 0 \leqslant t < n\} \leqslant \hat{y} \leqslant \max\{y_t : 0 \leqslant t < n\}.$$

With this definition, show that relation 6 holds approximately for any ω.

Exercise 6.4   The Impulse Response Function

Suppose that a linear filter has weights $\{ g_u : r \leqslant u \leqslant s \}$, where $r \leqslant 0 \leqslant s$, and that its input is

$$y_t = \begin{cases} 1 & \text{if } t = v \\ 0 & \text{otherwise} \end{cases}$$

for some v satisfying $s \leqslant v < n + r$. Show that the output satisfies

$$z_t = \begin{cases} g_{t-v} & \text{if } v + r \leqslant t \leqslant v + s, \\ 0 & \text{otherwise}. \end{cases}$$

6.3   DESIGNING A FILTER

There are basically two distinct ways of designing a filter for any given application. The first is to take a convenient set of primitive filters, such as the simple moving averages of various lengths, and use them as building blocks in assembling a filter with the desired characteristics. The second is to specify the requirements fairly precisely and then construct a filter directly to satisfy them. We shall give examples of both ways in this section and the next, and apply the resulting methods to complex demodulation in the remaining sections of this chapter.

The requirements for a filter are usually specified in terms of both the impulse and the frequency response functions. Other things being equal, it is usually desirable for the filter to have a short *span*, where the span is

$$\max\{u : g_u \neq 0\} - \min\{u : g_u \neq 0\}.$$

The span is one more than the total number of points that are lost (or have to be computed specially) at the ends of the data, and the computational effort required to apply the filter is usually roughly proportional to the span. It is usually undesirable for the impulse response function to show eccentric behavior such as large ripples.

The desired frequency response function, or transfer function, may usually be specified more precisely. The problem which arose in Section 6.1, that of separating a low-frequency component from other terms, is typical. A filter that does this is called a *low-pass* filter. Its transfer function should be close to 1 for frequencies in the interval associated with the low-frequency component (the *pass band*) and close to 0 elsewhere (the *stop band*). Since transfer functions are continuous, there is necessarily an intermediate band, the *transition band*, in which the transfer function lies between 0 and 1.

The transfer function of a symmetric filter (we shall use only symmetric filters) is a polynomial in $\cos\omega$, and hence these requirements cannot be met exactly for all frequencies. For complex demodulation we can use additional information to decide which frequencies are most important. From (4) we see that there is a term with frequencies around 2λ which should be removed. (Strictly the frequency is -2λ, but since the transfer function is symmetric the sign may be ignored.) Thus the transfer function should be 0 at 2λ, and small in a band surrounding 2λ. The other frequencies that need to be removed are those in the final term. Since $\{z_t\}$ is known to contain a low-frequency term and the second harmonic of λ, this term contains power at around λ and 3λ. Thus we similarly need zeros at λ and 3λ.

The simplest way to come close to satisfying these requirements is to find a polynomial $P(x)$ such that

$$P(1)=1, \qquad P(\cos\lambda)=P(\cos 2\lambda)=P(\cos 3\lambda)=0. \qquad (7)$$

The transfer function $P(\cos\omega)$ is necessarily close to 1 for low frequencies and close to 0 around λ, 2λ, and 3λ. The lowest degree polynomial satisfying (7) is

$$P(x)=\frac{(x-\cos\lambda)(x-\cos 2\lambda)(x-\cos 3\lambda)}{(1-\cos\lambda)(1-\cos 2\lambda)(1-\cos 3\lambda)}.$$

The graph of $P(\cos\omega)$ is shown in Figure 6.2 for $\lambda=0.565$, the approximate

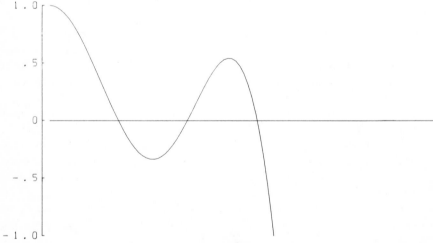

Figure 6.2   A transfer function with three zeros.

frequency of the sunspot cycle (in radians per year). The values between the zeros are unacceptably large, and the function becomes large and negative for frequencies much larger than $\pi/2$. Thus frequencies close to π would be strongly amplified (and have their phase changed by π). If we add further zeros at 4λ and 5λ, the transfer function is as in Figure 6.3, which is more acceptable.

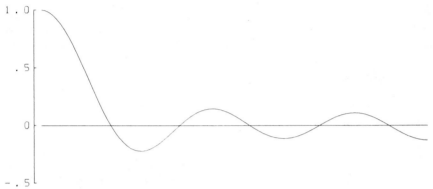

Figure 6.3   A transfer function with five zeros.

The filter corresponding to any transfer function can be found in two ways. The more direct is to replace $\cos\omega$ by $\frac{1}{2}\{\exp(i\omega)+\exp(-i\omega)\}$ and expand in powers of $\exp(i\omega)$. The impulse response function is given by the coefficients in this expansion. The second way is convenient when the transfer function is specified in a factorized form,

$$P(\cos\omega) = \prod_j (a_j + b_j\cos\omega).$$

The factor $a_j + b_j\cos\omega$ is the transfer function of the three-term filter with weights $b_j/2$, a_j, $b_j/2$. Thus, if we apply these three-term filters successively, the final series is the same as the output from the desired filter (see Section 6.2). This is an example of building a filter using three-term filters as primitives.

The simplest way to fill in the two end values when using a three-term filter is to assume that both y_{-1} and y_n vanish. A modification would be to multiply the values obtained in this way by $(a_j + b_j)/(a_j + b_j/2)$, thus making the sum of the two coefficients used here the same as the sum of the three coefficients used elsewhere. Either of these rules destroys the symmetry of the filter, however, and it will often be preferable just to multiply y_0 (or y_{n-1}) by $a_j + b_j$ and to use the value obtained.

When the frequency λ corresponds to an integer period p, the p-term *simple moving average* provides a useful filter. If p is odd, $p = 2q + 1$, this is

$$z_t = \frac{1}{p} \sum_{u=-q}^{q} y_{t-u}, \tag{8}$$

with transfer function $D_p(\omega)$, the Dirichlet kernel (see Section 2.2). For $D_p(0) = 1$ and $D_p(k\lambda) = 0$, $k = 1, 2, \ldots, q$. This transfer function therefore has zeros at the required places, and its oscillations become smaller as ω increases. For the sunspot data the period is close to 11 years, and in fact $D_{11}(\omega)$ is very similar to the transfer function shown in Figure 6.3 (see also Exercise 6.6). If p is even, $p = 2q$, there is no symmetric simple moving average of length p. The modified average

$$z_t = \frac{1}{p} \left(\tfrac{1}{2} y_{t-q} + \sum_{u=-q+1}^{q-1} y_{t-u} + \tfrac{1}{2} y_{t+q} \right) \tag{9}$$

is ordinarily used. It is the result of using an asymmetric simple moving average of length p followed by an asymmetric simple moving average of length 2, and hence its transfer function is $D_p(\omega) D_2(\omega) = \cos \omega / 2 D_p(\omega)$ (see Exercise 6.7). The main effect of the factor $\cos \omega / 2$ is to add a second zero at π.

For simple moving averages a natural rule for filling in end values is to take the average of the available data. For example, if $p = 2q + 1$ we could use the values

$$\frac{1}{q+1} \sum_{u=-q}^{0} y_{-u}, \qquad \frac{1}{q+2} \sum_{u=-q}^{1} y_{1-u}, \qquad \cdots$$

for $z_0, z_1, \ldots$. However, this has the unfortunate aspect that these averages are asymmetric. An alternative is to use *centered* averages of the available data. In this case we use y_0 for z_0, $\tfrac{1}{3}(y_0 + y_1 + y_2)$ for z_1, and so on up to z_q.

Since simple moving averages are easy to compute, we could add them to the set of primitive filters. A filter could then be built by starting with a suitable moving average (or combination of moving averages), and using three-term filters to impose any further zeros that might be needed in the transfer function. End values would be filled in by the appropriate rule at each step.

The transfer function may be made arbitrarily small in the stop band by these techniques. However, we have done nothing to improve its behavior in the pass band. If we decide that the transfer function $G(\omega)$ decreases too rapidly as ω increases from 0, the following simple modification provides an improvement. Suppose that $\{y_t\}$ is the input, $\{z_t^{(1)}\}$ is the output, and $\{z_t^{(2)}\}$ is the result of a second application of the filter (that is, the output when $\{z_t^{(1)}\}$ is the input). Let $z_t = 2z_t^{(1)} - z_t^{(2)}$. It may be shown that $\{z_t\}$ is the output of a new filter, whose transfer function is $G_2(\omega) = 2G(\omega) - G(\omega)^2 = 1 - \{1 - G(\omega)\}^2$. Thus $G_2(\omega)$ stays closer to 1 than does $G(\omega)$ but has zeros at the same frequencies. Hence we have widened the pass band without seriously affecting the stop band. [Note that in the stop band $|G(\omega)| \ll 1$ and thus $G_2(\omega) \cong 2G(\omega)$.] The cost is that the span of the filter is nearly doubled (see Exercise 6.5).

Exercise 6.5   Combination of Filters

Suppose that a filter is obtained by using a filter with weights $\{ g_u : r \leqslant u \leqslant s \}$ on the output of another with weights $\{ g'_u : r' \leqslant u \leqslant s' \}$. Show that the span of the combined filter is the sum of the individual spans less 1, and find the impulse response function of the combined filter.

Exercise 6.6   The Impulse Response Function

Find the impulse response function of the filter whose transfer function has zeros at $j\lambda, j = 1, \ldots, 5$, for $\lambda = 0.565$, and has the value 1 at $\omega = 0$. Verify that each weight is approximately $1/11$.

Exercise 6.7   Simple Moving Averages of Even Length

Verify that the modified simple moving average of length $p = 2q$, equation 9, gives the same output as the combined asymmetric filters

$$z'_t = \frac{1}{p} \sum_{u = -q+1}^{q} y_{t-u},$$

$$z_t = \tfrac{1}{2}(z'_t + z'_{t+1}).$$

6.4   LEAST SQUARES FILTER DESIGN

A more systematic way of designing a filter is as follows. Suppose that we have an ideal transfer function $H(\omega)$ and that we decide to approximate it by a finite impulse response filter. For instance, to design a low-pass filter we might decide on a *pass frequency* ω_p, a *stop frequency* ω_s, and *tolerances* δ_p and δ_s, and then look for the filter with the shortest span whose transfer function $G(\omega)$ satisfies

$$|1 - G(\omega)| \le \delta_p, \qquad 0 \le \omega \le \omega_p,$$

$$|G(\omega)| \le \delta_s, \qquad \omega_s \le \omega \le \pi.$$

Such problems are usually complex, and can be solved only by numerical optimization.

An easier approach is to fix the range of the impulse response function and then find the filter whose transfer function best approximates $H(\omega)$ in the least squares sense. In other words, for given r and s we find $\{g_u : r \le u \le s\}$ to minimize

$$\int_{-\pi}^{\pi} |H(\omega) - \sum_{u=r}^{s} g_u \exp(-iu\omega)|^2 \, d\omega. \tag{10}$$

It may be verified that the solution is

$$g_u = h_u = \frac{1}{2\pi} \int_{-\pi}^{\pi} H(\omega) \exp(iu\omega) \, d\omega, \qquad u = r, \dots, s, \tag{11}$$

the *Fourier coefficients* of $H(\omega)$ (see Exercise 6.8). It is interesting and very convenient that the optimal value of g_u does not depend on the values of r and s. The approximate transfer function is

$$H_{r,s}(\omega) = \sum_{u=r}^{s} h_u \exp(-iu\omega)$$

$$= \frac{s-r+1}{2\pi} \int_{-\pi}^{\pi} H(\lambda) D_{s-r+1}(\omega - \lambda) \exp\left\{ \frac{-i(r+s)(\omega-\lambda)}{2} \right\} d\lambda \tag{12}$$

(see Exercise 6.9), the Fourier series of H truncated at r and s. If H is real and symmetric and we set $r = -s$, we find that

$$h_u = \frac{1}{2\pi} \int_{-\pi}^{\pi} H(\omega) \cos u\omega \, d\omega$$

$$= \frac{1}{\pi} \int_{0}^{\pi} H(\omega) \cos u\omega \, d\omega = h_{-u}$$

and

$$H_{-s,s}(\omega) = \sum_{u=-s}^{s} h_u \exp(-iu\omega)$$

$$= h_0 + 2 \sum_{u=1}^{s} h_u \cos u\omega$$

$$= \frac{2s+1}{2\pi} \int_{-\pi}^{\pi} H(\lambda) D_{2s+1}(\omega - \lambda) d\lambda$$

$$= H_s(\omega), \qquad \text{say.} \tag{13}$$

Suppose that we use this method to approximate the ideal low-pass transfer function

$$H(\omega) = \begin{cases} 1 & \text{if } 0 \leqslant \omega \leqslant \omega_c, \\ 0 & \text{if } \omega_c < \omega \leqslant \pi, \end{cases} \tag{14}$$

where ω_c is the *cutoff frequency*. The Fourier coefficients are

$$h_u = \frac{1}{\pi} \int_0^{\omega_c} \cos u\omega \, d\omega$$

$$= \frac{\sin u\omega_c}{\pi u}, \qquad u \geqslant 1,$$

and

$$h_0 = \frac{\omega_c}{\pi} .$$

For example, in the sunspot data a reasonable cutoff frequency would be $\omega_c = \lambda/2 = 0.283$. Figure 6.4 shows the approximating transfer functions for $s = 5$ (an 11-term filter) and $s = 20$ (a 41-term filter), and the ideal transfer function. The latter curve shows pronounced *overshoot* on either side of the cutoff frequency. This is known as *Gibbs's phenomenon* (Lanczos, 1961, p. 225) and is characteristic of the truncated Fourier series of a discontinuous function.

This overshoot and the accompanying ripples may be greatly reduced as follows. It may be shown that the wavelength of the ripples is $\delta = 4\pi/(2s+$

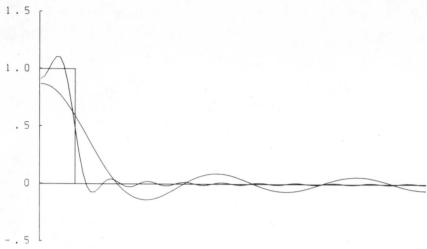

Figure 6.4   Transfer functions of least squares low-pass filters, $s = 5$ and $s = 20$.

1) (see Exercise 6.10). Thus the smoothed function

$$\tilde{H}_s(\omega) = \frac{1}{\delta} \int_{\omega - \delta/2}^{\omega + \delta/2} H_s(\lambda) \, d\lambda$$

has far smaller ripples, since the integration is over one complete cycle. But

$$\tilde{H}_s(\omega) = h_0 + 2 \sum_{u=1}^{s} h_u \frac{\sin u\delta/2}{u\delta/2} \cos u\omega, \qquad (15)$$

and this corresponds to replacing the Fourier coefficient h_u by

$$h_u \frac{\sin u\delta/2}{u\delta/2} = h_u \frac{\sin 2\pi u/(2s+1)}{2\pi u/(2s+1)}. \qquad (16)$$

The multipliers

$$\sigma_{s,u} = \frac{\sin 2\pi u/(2s+1)}{2\pi u/(2s+1)}$$

are an example of *convergence factors* and are essentially the same as the σ-factors introduced by Lanczos (1961, pp. 225–229) to accelerate the convergence of Fourier series. Figure 6.5 shows the functions $\tilde{H}_5(\omega)$ and

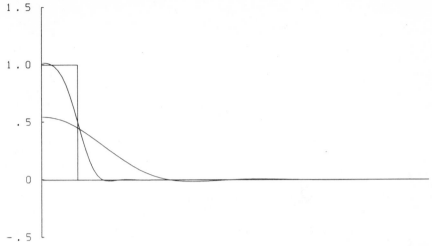

Figure 6.5   Transfer functions of least squares low-pass filters with convergence factors applied, $s = 5$ **and** $s = 20$.

$\tilde{H}_{20}(\omega)$ corresponding to the functions in Figure 6.4. The ripples in each have been substantially reduced, as has the overshoot in $\tilde{H}_{20}(\omega)$. The absolute values of these functions are plotted on a logarithmic scale in Figure 6.6. The plot shows that the sidelobes, especially those of $\tilde{H}_{20}(\omega)$, are uniformly small. (The troughs between the sidelobes actually extend to $-\infty$. They show finite local minima because the plot was prepared by linear interpolation of function values on a discrete grid.)

The use of convergence factors is analogous to the use of a data window (see Section 5.2). A truncated Fourier series may be regarded as the infinite series with the coefficients multiplied by a boxcar function. This multiplication is equivalent to convolving the original function with the transform of the boxcar, the Dirichlet kernel. We have in fact derived this equivalence in (12). The convergence factors, which are initially 1 and decay smoothly to 0, are a smooth approximation to the boxcar. The modified truncated sum can be represented as the convolution of the original function with the transform of the convergence factors (see Exercise 6.11), which has smaller sidelobes than the Dirichlet kernel. A split cosine bell taper (Section 5.2) would have achieved much the same effect. Note that the cosine bell in fact approaches 0 more smoothly than the convergence factors, which behave like $(\sin x)/x$ (see Figure 3.1). It is therefore likely that cosine bell tapering applied only to the higher coefficients would be roughly equivalent to the use of the factors derived in this section.

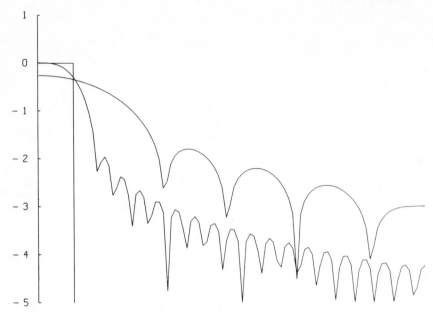

Figure 6.6   Absolute values of transfer functions in Figure 6.5, logarithmic vertical scale.

Another interpretation of the application of convergence factors may be derived as follows. The convergence factors are the Fourier coefficients of the (continuous) boxcar

$$C(\lambda) = \begin{cases} 2\pi/\delta & \text{if } |\lambda| \leq \delta/2, \\ 0 & \text{otherwise,} \end{cases}$$

and therefore the products (16) are the Fourier coefficients of the convolution of the ideal transfer function $H(\omega)$ with this boxcar (see Exercise 6.12). Thus the modified partial sum (15) may also be regarded as the least squares approximation to this convolution, which is a smoothed version of the ideal transfer function.

This gives us a dual interpretation of the modified transfer function, as

(i)   a smoothed version of the least squares approximation to the ideal transfer function (the original derivation), or

(ii)   the least squares approximation to a smoothed version of the ideal transfer function.

For the ideal low-pass filter (14), the effect of this smoothing is to replace the ideal low-pass transfer function, with its sharp cutoff at ω_c, by a modified function that decays linearly from the value 1 at $\omega_c - \delta/2$ to 0 at

$\omega_c + \delta/2$. Thus we have introduced a *transition band* of width $\delta = 4\pi/(2s + 1)$. Actually, this is true only if $\omega_c \geqslant \delta/2$. In the case of $\tilde{H}_5(\omega)$, we have $\delta/2 = 2\pi/11 \cong 0.571$, while $\omega_c = 0.283$. If $\omega_c < \delta/2$, we actually have a transition band that extends from $\delta/2 - \omega_c$ to $\delta/2 + \omega_c$. Thus, for a given value of ω_c, we should avoid the use of convergence factors if $\omega_c < \delta/2 = 2\pi/(2s + 1)$, that is, if $2s + 1 < 2\pi/\omega_c$. This explains the poor approximation given by $\tilde{H}_5(\omega)$. In the present case the value $2\pi/\omega_c$ is around 22. The ideal function, the transition band version for $s = 20$, and $\tilde{H}_{20}(\omega)$ are shown in greater detail in Figure 6.7 (for $0 \leqslant \omega \leqslant \pi/4$).

It is usually desirable for a filter to pass a zero-frequency component (that is, a constant term) without change. This requires that the transfer function have the value 1 at zero frequency. For the filters constructed in Section 6.3 this requirement is easily imposed, but Figures 6.4 to 6.6 show that least squares approximations to an ideal filter with this property do not in general share it, whether or not convergence factors are used. The least squares argument is easily modified to include this constraint, the only effect on the solution being that h_u is replaced by $h_u + (1 - \Sigma_{v=r}^{s} h_v)/(2s + 1)$. When convergence factors $\{\sigma_{s,u} : -s \leqslant u \leqslant s\}$ are used, an appropriate modification is to replace h_u by

$$h_u + \frac{1 - \sum_v h_v \sigma_{sv}}{\sum_v \sigma_{sv}}.$$

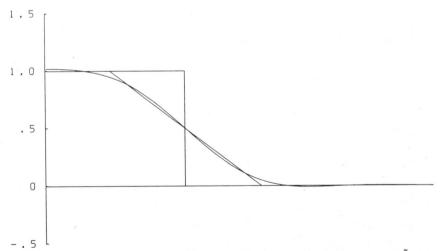

Figure 6.7   Ideal low-pass transition functions with and without transition bands, and $\tilde{H}_{20}(\omega)$.

In either case, a simple alternative is just to rescale the coefficients so that they sum to 1.

Filling in end values is not as easy with the filters described in this section as with the three-term and simple moving average filters of the preceding section. The least squares approach leads us to constrain the weights attached to unavailable data to be 0 and then to choose the remaining coefficients optimally. However, these optimal values are not affected by the constraints, and thus the overall effect is the same as replacing unavailable values by 0. The resulting filter becomes quite asymmetric at the extreme ends of the series and hence may be unacceptable in some cases. If convergence factors are used, they should be computed separately for the two sides of the filter. Thus for $t < s$, for instance, we would compute the output as

$$\sum_{u=0}^{t} h_u \sigma_{t,u} y_{t-u} + \sum_{u=-s}^{-1} h_u \sigma_{s,u} y_{t-u}.$$

Exercise 6.8   The Fourier Coefficients

Verify that the values of g_u, $u = r, \ldots, s$, that minimize (10) are the Fourier coefficients (11). [*Hint*: Write $H(\omega)$ as $X(\omega) + iY(\omega)$ and g_u as $x_u + iy_u$, expand (10) in terms of these real quantities, and differentiate with respect to x_u and y_u.]

Exercise 6.9   The Truncated Fourier Series

Verify (12). [*Hint*: Substitute the integral formula (11) for h_u, and sum.]

Exercise 6.10   Gibbs's Phenomenon

Simplify (13) in the case of the ideal low-pass filter (14). Show that, if s is large and ω_c is small, the oscillations in $H_s(\omega)$ in any small interval (ω_1, ω_2) not containing 0 are approximately sinusoidal, with period $4\pi/(2s+1)$. [*Hint*: The denominator of $D_n(\lambda)$ is approximately constant in any short interval.]

Exercise 6.11   General Convergence Factors

Suppose that $\{h_u\}$ are the Fourier coefficients of $H(\omega)$, defined as in (11), and that $\{c_u : r \leqslant u \leqslant s\}$, $r \leqslant 0 \leqslant s$, are a set of numbers, which in the present context are to be interpreted as convergence factors. Let

$$H_c(\omega) = \sum_{u=r}^{s} c_u h_u \exp(-iu\omega)$$

be the corresponding modified partial sum of the Fourier series $H(\omega)$. Show that

$$H_c(\omega) = \int_{-\pi}^{\pi} H(\lambda) C(\omega - \lambda) d\lambda,$$

where

$$C(\lambda) = \frac{1}{2\pi} \sum_{u=r}^{s} c_u \exp(-iu\lambda).$$

[*Hint*: Substitute the integral formula for h_u.]

NOTE 1:   We have made no assumption about the convergence of the infinite Fourier series of $H(\omega)$.

NOTE 2:   This identity is analogous to the discrete result derived in Exercise 5.4. In words, we have shown that *the Fourier coefficients of the convolution of two functions are the products of the respective Fourier coefficients*, provided that one of the functions has a finite Fourier series [in this case, $C(\lambda)$].

Exercise 6.12   (Continuation) A Generalization

Suppose that $H_c(\omega)$ is the convolution of $H(\omega)$ and $C(\omega)$, as in Exercise 6.11, but make no assumption about the finiteness or convergence of the Fourier series of either $H(\omega)$ or $C(\omega)$. Show that the Fourier coefficients of $H_c(\omega)$ are the products of those of $H(\omega)$ and $C(\omega)$.

6.5   DEMODULATING THE SUNSPOT SERIES

For the first example of complex demodulation we use the sunspot data introduced in Section 5.4. It was verified in that section that the oscillations in these data have a period of around 11 years, but that the amplitude and the phase of the oscillations tend to vary. Figure 6.8 shows the *instantaneous amplitude* and *instantaneous phase* of these oscillations as functions of time. These were calculated by forming the demodulated series

$$y_t = x_t \exp(-i\lambda t), \qquad t = 0, \ldots, n-1,$$

where x_t is the tth annual sunspot number (corresponding to the year $1700 + t$), and $\lambda = 2\pi/11$ is the approximate frequency. This series was then smoothed by taking a simple moving average of 11-year blocks. As is explained in Section 6.3, this removes most of the unwanted components

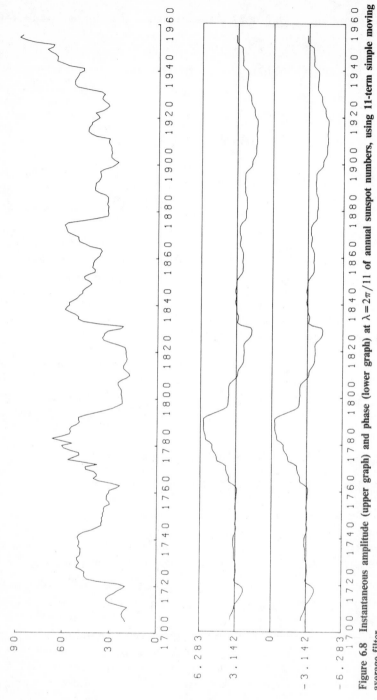

Figure 6.8 Instantaneous amplitude (upper graph) and phase (lower graph) at $\lambda = 2\pi/11$ of annual sunspot numbers, using 11-term simple moving average filter.

of $\{y_t\}$ (see equation 4). The integer period of 11 years was used in preference to the period of 11.13 years, which corresponds to the largest peak in the periodogram, because the former value allows us to use a simple moving average. With this combination of frequency and filter, the interpretation of complex demodulation as a local version of harmonic analysis, mentioned in Section 6.1, is in fact an exact description, for the tth term in the smoothed demodulated series is just the discrete Fourier transform of $x_{t-5}, \ldots, x_{t+5}$ evaluated at $2\pi/11$, the first Fourier frequency for a series (or subseries) of length 11. In the same way, if a filter with nonconstant weights is used, the tth term in the demodulated series is the transform of a *tapered* stretch of data surrounding time t, the data window weights being (proportional to) the weights of the filter.

The smoothed series is

$$z_t = u_t + iv_t \cong \tfrac{1}{2} R_t \exp(i\phi_t),$$

and we estimate R_t and ϕ_t by solving the equations $R\cos\phi = 2u_t$, $R\sin\phi = 2v_t$. The solution for R is $2(u_t^2 + v_t^2)^{1/2}$, while for ϕ it is given by the modification of arctan v_t/u_t described in Section 2.2 [computed in FOR-TRAN as ATAN2(v_t, u_t)].

The amplitude plot (Figure 6.8) shows that there are indeed substantial variations in the amplitude of the oscillations, with a range of about $3:1$. The raggedness of this graph shows that the 11-year moving average did not remove all of the unwanted components. The short-term fluctuations in the graph cannot be interpreted as fluctuations in the amplitude of a sinusoid with a period of 11 years.

The phase plot also displays substantial variations. The graph shows the phase in each year both as its principal value (which is defined to lie in the range $-\pi$ to π) and as the principal value $\pm 2\pi$, with the sign chosen to make this second value lie in the range -2π to 2π. In this way we have a continuous curve to examine even where the principal value jumps from π to $-\pi$.

Note that in any time interval in which the phase is a linear function $a + bt$ the oscillations are generated by

$$R_t \cos(\lambda t + a + bt) = R_t \cos\{(\lambda + b)t + a\},$$

a modulated sinusoid of frequency $\lambda + b$. Overall, the the graph shows a slight downward slope, which indicates that the basic frequency is slightly below $2\pi/11$, or in other words that the basic period is slightly greater than 11 years. However, by estimating this slope from different stretches of data we could arrive at very different answers, and we conclude that the period is not well determined by these data.

Figure 6.9 shows the instantaneous amplitude and phase calculated using a more sophisticated filter in place of the 11-year moving average. The filter is a 41-term least squares approximation to a low-pass filter with cutoff at $\lambda/2 = 2\pi/22$, using convergence factors. It is thus very like the filter whose tranfer function is shown in Figures 6.5 and 6.6. The much smaller sidelobes in this transfer function remove more of the unwanted components, and hence the two graphs, especially the amplitude plot, are smoother. The fact that broad features of these graphs are very similar to those of the graphs in Figure 6.8 suggests that the extra smoothness has not been gained at the cost of loss of accuracy.

The end values have not been filled in for either of the filters used to obtain Figures 6.8 and 6.9. The procedures described in Sections 6.3 and 6.4 for filling in end values either are asymmetric, thus introducing biases into the smoothed values, or involve only few data points, leaving them insufficiently smoothed. Both of these aspects are undesirable, and unless the end values are important, it may be preferable not to compute them. In the case of the 11-year simple moving average this amounts to losing 5 points at each of the data, which usually will not be important. With the 41-term filter, however, we lose 20 points at each end, which may well be important, particularly if we are interested in forecasting the amplitude and phase of future oscillations.

Figures 6.8 and 6.9 give us information about the variations in the fundamental frequency of the oscillations. If we are interested in the nonsinusoidal nature of the oscillations, we may calculate the *instantaneous relative phase* of the second harmonic. This is the local analog of the relative phase defined in Section 5.6 and is computed by complex demodulation at frequency 2λ as well as λ. The upper curve in Figure 6.10 shows the result when the 11-year simple moving average is used at both frequencies, while the lower curve results from using the 41-term filter. In each case the quantity plotted is $\phi_{2t} - 2\phi_{1t}$, where ϕ_{1t} and ϕ_{2t} are the instantaneous phases at time t and frequencies $\lambda = 2\pi/11$ and $2\lambda = 4\pi/11$, respectively. As before, the principal value and one other are plotted for each t. Both curves lie largely in the range 0 to $\pi/2$, as is to be expected from the nature of the oscillations (see Section 5.6). The main departures are fairly sharp jumps of 2π, visible five times in the upper graph and twice in the lower. These occur where the amplitude of the second harmonic is not strong, and thus its phase is not well determined.

6.6 ANOTHER EXAMPLE OF COMPLEX DEMODULATION

The second set of data we use to illustrate complex demodulation is shown in Figure 6.11. They are 3000 observations of the disturbance in the

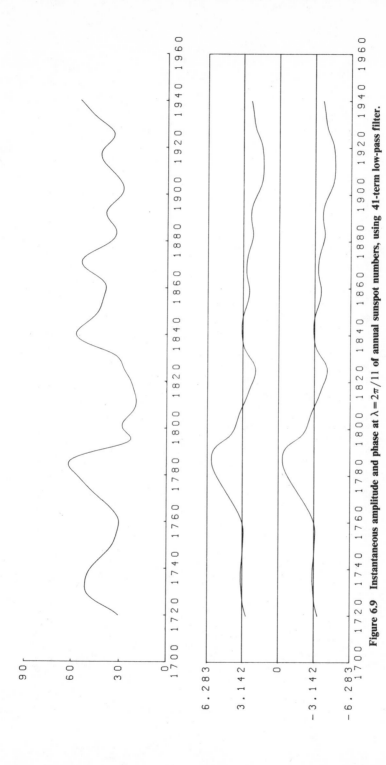

Figure 6.9 Instantaneous amplitude and phase at $\lambda = 2\pi/11$ of annual sunspot numbers, using 41-term low-pass filter.

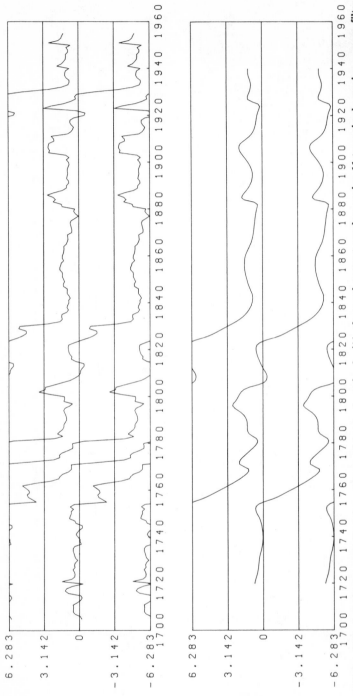

Figure 6.10 Instantaneous relative phase of second harmonic at $\lambda = 2\pi/11$ of annual sunspot numbers, using 11-term simple moving average filter (upper graph) and 41-term low-pass filter (lower graph).

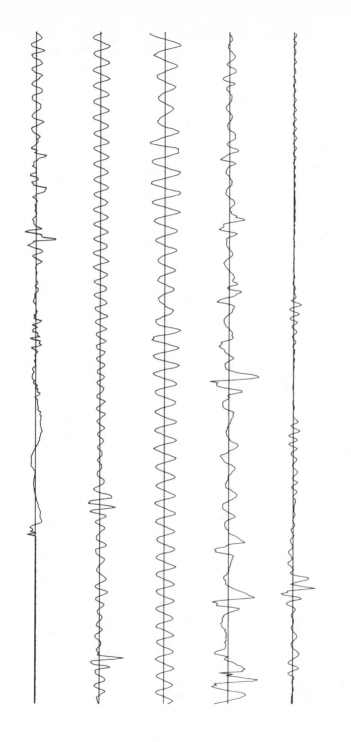

Figure 6.11 Three thousand observations on a plasma (shown in six segments, each containing 500 observations).

143

magnetic field of plasma in a TOKAMAK plasma generator. The sampling interval is 10 microseconds, so that the entire span of the data covers 30 milliseconds. The most striking feature of the data is the buildup of a sinusoidal oscillation that suddenly breaks down into ragged fluctuations. The sinusoidal oscillation then gradually re-emerges toward the end of the experiment.

Figure 6.12 shows the periodogram of $x_{1250}, \ldots, x_{1505}$ (256 terms, tapered 20%), which is the stretch of data immediately preceding the breakdown. The largest ordinate is the 17th, whose frequency is $\omega_{17} = 2\pi \times 17/256 = 0.42$. Adjacent values are also large because of the amplitude and phase fluctuations visible in the original graph. There is also some slight evidence of the presence of the second harmonic, in that a few ordinates around the 34th are a little larger than would be expected from the general rate of decay. With 20% tapering, these effects cannot be due to leakage from the main peak.

When we complex-demodulate these data, we want the whole main peak to show in the demodulated series. Since the peak extends roughly from the 12th to the 24th ordinate, we should demodulate at the frequency $\lambda = \omega_{18} = 2\pi \times 18/256$ (the middle of this band), and the filter that we use to remove unwanted components should pass all frequencies out to $\omega_6 = 2\pi \times 6/256$ (one-half the width of the band). On the other hand, the terms that could be second harmonics need to be removed, and they begin to show at around $\omega_{32} = 2\pi \times 32/256$. Demodulation moves these components down to frequency $\omega_{14} = 2\pi \times 14/256$, and thus our filter should stop all frequencies from this point on.

We can use the theory of Section 6.4 to find such a filter. It was shown that, if convergence factors are applied to the $(2s + 1)$-term least squares approximation to the ideal low-pass filter with cutoff frequency ω_c, the result is a low-pass filter with a transition band extending from $\omega_c - \delta/2$ to

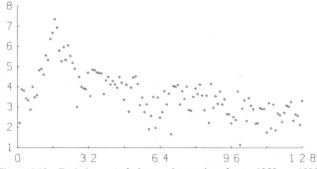

Figure 6.12   Periodogram of plasma observations for $t = 1250, \ldots, 1505$.

$\omega_c + \delta/2$, where $\delta = 4\pi(2s + 1)$. Thus we solve the equations

$$\omega_c - \frac{\delta}{2} = 2\pi \times \frac{6}{256},$$

$$\omega_c + \frac{\delta}{2} = 2\pi \times \frac{14}{256},$$

finding $\omega_c = 2\pi \times 10/256$ and $\delta = 2\pi \times 8/256$. The choice $s = 32$, giving a 65-term filter, is convenient.

Figure 6.13 shows for $t = 1000$ to 2000 the instantaneous amplitude and phase that result from demodulating at $\omega_{18} = 2\pi \times 18/256$ and using this filter. Both the amplitude plot and the phase plot reflect clearly the change in behavior at around $t = 1500$. Before this point the roughly parabolic shape of the phase plot indicates that the frequency was slowing down more or less as a linear function of time. In the same interval the amplitude shows three rises, to successively higher peaks, each followed by a sharp drop. Each peak is associated with a noticeable increase in frequency (indicated by a steeper positive or less steep negative slope in the phase plot), and the troughs following the first two peaks are associated with noticeable decreases. Since these changes in frequency are very short-lived, we could alternatively say that the cycles with the largest amplitudes occur a little early, whereas those following occur a little late. The erratic behavior of both curves in the second half of the interval, from 1500 to 2000, indicates that the data contain no consistent component near the demodulation frequency in this half.

The data shown in Figure 6.11 are the first of seven series of observations made simultaneously on the plasma. Using all seven series, we may extract more information about the behavior of the plasma, using the multiple series methods described in Chapter 9.

APPENDIX

The following program was used to carry out some of the complex demodulations described in this chapter. Subprogram LOPASS applies the least squares low-pass filter with convergence factors (see Section 6.4). Subprogram SIMPMA (included here, but not used by the current program) was used for the simple moving average filtering, by making a minor modification to the main program.

NOTE: The program also uses subprograms DATIN (given in the Appendix to Chapter 2), DETRND, and DATOUT (in the Appendix to Chapter 5).

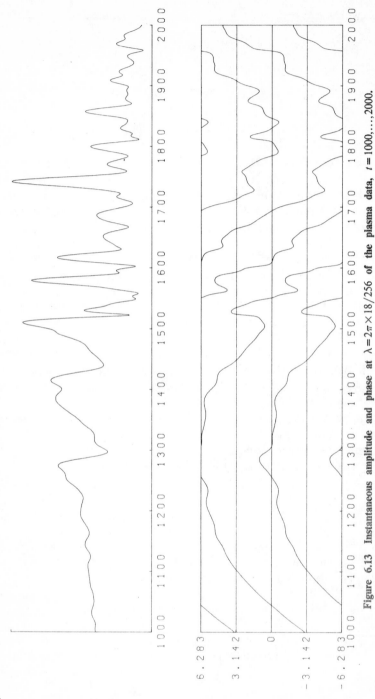

Figure 6.13 Instantaneous amplitude and phase at $\lambda = 2\pi \times 18/256$ of the plasma data, $t = 1000, \ldots, 2000$.

146

```
C
C     THIS PROGRAM DEMODULATES SERIES X AT FREQUENCY OMEGA.
C     THE DEMODULATED SERIES IS THEN FILTERED WITH A LEAST-
C     SQUARES LOPASS FILTER USING CONVERGENCE FACTORS.
C     THE PROGRAM IS CONTROLLED BY THE FOLLOWING VARIABLES,
C     WHICH ARE INPUT TO THE PROGRAM -
C
C     NAME     VALUE
C
C     NP2      A POWER OF 2
C
C     NOM      (NP2/2*PI)*OMEGA
C
C     NPA      (NP2/2*PI) TIMES THE PASS FREQUENCY OF THE FILTER
C
C     NST      (NP2/2*PI) TIMES THE STOP FREQUENCY OF THE FILTER
C
      DIMENSION X(1080),D1(1080),D2(1080)
      DATA PI /3.141593/
      CALL DATIN (X,N,START,STEP)
      CALL DETRND (X,N,0)
      READ (5,4) NP2,NOM,NPA,NST
    4 FORMAT(4I5)
      WRITE(6,5) NOM,NP2,NPA,NP2,NST,NP2
    5 FORMAT(≠0COMPLEX DEMODULATION AT 2*PI*≠,I5,≠/≠,I5/
     +        ≠ PASS FREQUENCY IS        2*PI*≠,I5,≠/≠,I5/
     +        ≠ STOP FREQUENCY IS        2*PI*≠,I5,≠/≠,I5)
      OMEGA=2.0*PI*FLOAT(NOM)/FLOAT(NP2)
      CUTOFF=PI*FLOAT(NPA+NST)/FLOAT(NP2)
      NS=NP2/(NST-NPA)
      WRITE(6,6) OMEGA,CUTOFF,NS
    6 FORMAT(≠0DEMODULATION FREQUENCY IS≠,F10.5/
     +        ≠ CUTOFF FREQUENCY IS        ≠,F10.5/
     +        ≠ THE FILTER USES 2*≠,I5,≠ + 1 TERMS≠)
      CALL DEMOD (X,N,OMEGA,D1,D2)
      CALL LOPASS (D1,N,CUTOFF,NS)
      CALL LOPASS (D2,N,CUTOFF,NS)
      LIM=N-2*NS
      CALL POLAR (D1,D2,LIM)
      START=START+FLOAT(NS)*STEP
      CALL DATOUT (D1,LIM,START,STEP)
      CALL DATOUT (D2,LIM,START,STEP)
      STOP
      END
```

```
      SUBROUTINE DEMOD (X,N,OMEGA,D1,D2)
C
C   THIS SUBROUTINE DEMODULATES THE SERIES X AT FREQUENCY
C   OMEGA.    THE REAL AND IMAGINARY PARTS OF THE DEMODULATED
C   SERIES ARE RETURNED IN D1 AND D2,   RESPECTIVELY.
C
      DIMENSION X(N),D1(N),D2(N)
      DO 10 I=1,N
      ARG=FLOAT(I-1)*OMEGA
      D1(I)=X(I)*COS(ARG)*2.0
  10  D2(I)=-X(I)*SIN(ARG)*2.0
      RETURN
      END




      SUBROUTINE SIMPMA (X,N,LEN)
C
C   THIS SUBROUTINE CARRIES OUT SIMPLE MOVING AVERAGING ON
C   THE SERIES   X.    EACH TERM IN THE SERIES IS REPLACED
C   BY THE AVERAGE OF ITSELF AND THE SUCCEEDING   (LEN - 1)
C   TERMS.    THE AVERAGING IS DONE IN PLACE.
C
      DIMENSION X(N)
      LENM1=LEN-1
      LIM=N-LENM1
      CON=1.0/FLOAT(LEN)
      DO 10 I=1,LIM
      DO 20 J=1,LENM1
  20  X(I)=X(I)+X(I+J)
  10  X(I)=X(I)*CON
      RETURN
      END
```

```
      SUBROUTINE LOPASS (X,N,OMEGA,NS)
C
C     THIS SUBROUTINE CARRIES OUT LOW-PASS FILTERING OF THE
C     SERIES  X.    THE FILTER IS THE (2 * NS + 1)-TERM
C     LEAST SQUARES APPROXIMATION TO THE CUTOFF FILTER
C     WITH CUTOFF FREQUENCY  OMEGA.    ITS TRANSFER FUNCTION
C     HAS A TRANSITION BAND OF WIDTH DELTA SURROUNDING OMEGA,
C     WHERE DELTA = 4*PI/(2*NS+1).    THE FILTERING IS DONE
C     IN PLACE.
C
      DIMENSION X(N),H(100)
      DATA PI /3.141593/
      D(Z)=SIN(Z)/Z
      NS1=NS-1
      NS2=2*NS
      LIM=N-NS2
      HO=OMEGA/PI
      CON=2.0*PI/FLOAT(NS2+1)
      SUM=HO
      DO 10 I=1,NS
      H(I)=HO*D(FLOAT(I)*OMEGA)*D(FLOAT(I)*CON)
      SUM=SUM+2.0*H(I)
   10 CONTINUE
      HO=HO/SUM
      DO 15 I=1,NS
   15 H(I)=H(I)/SUM
      DO 20 I=1,LIM
      TEMP=HO*X(I+NS)
      DO 30 J=1,NS
   30 TEMP=TEMP+(X(I+NS+J)+X(I+NS-J))*H(J)
   20 X(I)=TEMP
      RETURN
      END
```

```fortran
      SUBROUTINE POLAR (X,Y,N)
C
C     THIS SUBROUTINE CONVERTS THE PAIR OF SERIES  X   AND   Y
C     FROM THE REAL AND IMAGINARY PARTS OF A SERIES OF COMPLEX
C     NUMBERS TO THEIR MAGNITUDES AND PHASES.   THE CONVERSION
C     IS DONE IN PLACE.   PARAMETERS ARE
C
C     NAME   TYPE                        VALUE
C                          ON ENTRY            ON RETURN
C
C     X      REAL ARRAY REAL PART            MAGNITUDE
C
C     Y      REAL ARRAY IMAGINARY PART       PHASE
C
C     N      INTEGER    SERIES LENGTH        UNCHANGED
C
      DIMENSION X(N),Y(N)
      DO 10 I=1,N
      R=SQRT(X(I)**2+Y(I)**2)
      PHI=ATAN2(Y(I),X(I))
      X(I)=R
   10 Y(I)=PHI
      RETURN
      END
```

7

THE SPECTRUM

We have seen that the results of harmonic analysis can be difficult to interpret even when the data show definite periodicity in the form of successive, fairly regular peaks and troughs. The sunspot series analyzed in Chapters 5 and 6 is ample evidence of this fact.

What, then, can we hope to achieve by harmonic analysis of a series with less well-defined oscillations, such as the Beveridge wheat-price series (Figure 1.3)? These data were collected and published by Beveridge (1921) as part of a study of the impact of meteorological variables on economic conditions. They are a price index for wheat in Europe, normalized so that the average price for 1700 to 1745 is 100. Historical records from 48 separate locations were combined to produce the index. Although the graph of the data shows a succession of peaks and troughs, these are by no means as regular as those of the sunspots. Nevertheless, harmonic analysis (and its close relative, *spectrum analysis*) of such economic series is widely used. We shall see in this chapter what kind of information it can yield.

7.1 PERIODOGRAM ANALYSIS OF WHEAT PRICES

The first problem in analyzing these data is the change in scale. No sinusoid can match oscillations that grow in amplitude. Beveridge produced an "index of fluctuation," shown in Figure 7.1, by dividing each value in the series by a centered average of 31 adjacent values. The oscillations in this index are more uniform, and in particular, show no tendency to change their amplitude over time. Beveridge (1921) gave various values of the periodogram of the index of fluctuations and in a later paper (1922) presented a more extensive analysis, including some corrections of the earlier values. The construction of the index of fluctuations may be motivated by modeling the data as $x_t = T_t I_t$, where T_t is the *trend* at time t and I_t is an *irregular* or oscillating term. The interpretation of these terms is that the trend reflects long-term economic forces such as inflation, whereas the irregular terms are caused by short-term effects such as fluctuations in supply from year to year. The wheat-price series was constructed to allow examination of these short-term fluctuations, and hence the trend term is an unwanted complication. We may regard the 31-year moving average as an approximation to the trend term, and thus the index of fluctuations is the corresponding estimate of the irregular component.

However, operations such as these introduce their own effects into the data, as is most easily seen in the case of the additive model $x_t = T_t + I_t$. It is natural to estimate T_t by applying a linear filter (see Section 6.2) to the data $\{x_t\}$. Suppose that the filter has weights $\{g_u: -s \leqslant u \leqslant s\}$ and transfer function $G(\omega)$. Then we approximate T_t by

$$y_t = \sum_{u=-s}^{s} g_u x_{t-u}. \tag{1}$$

Now $\{T_t\}$ is assumed to be smooth, and thus $T_{t-u} \cong T_t$ for $-s \leqslant u \leqslant s$ (provided that s is reasonably small). Hence

$$y_t = \sum g_u T_{t-u} + \sum g_u I_{t-u}$$

$$\cong T_t + \sum g_u I_{t-u}$$

provided that $\sum g_u = 1$ (which is the natural normalization). We approximate I_t by

$$z_t = x_t - y_t$$

$$\cong I_t - \sum g_u I_{t-u},$$

which is the result of applying a linear filter to $\{I_t\}$. The transfer function of the filter is

$$1 - \sum g_u \exp(-iu\omega) = 1 - G(\omega).$$

Thus the series $\{z_t\}$ is really an approximation to a filtered version of $\{I_t\}$. In the case of the multiplicative model $x_t = T_t I_t$ the trend term is

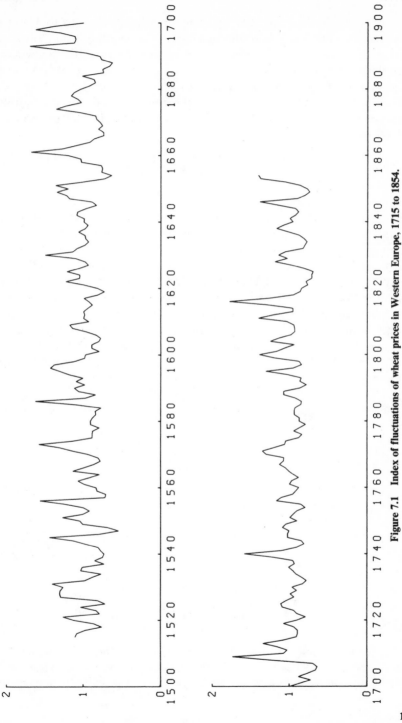

Figure 7.1   Index of fluctuations of wheat prices in Western Europe, 1715 to 1854.

approximated in the same way by applying a linear filter to $\{x_t\}$, but it is removed by division. The resulting series $z_t = x_t/y_t$, the general analog of Beveridge's index of fluctuations, is also approximately a filtered version of $\{I_t\}$, with the same transfer function $1 - G(\omega)$, as is shown by Granger and Hughes (1971) (see Exercise 7.1). For the index of fluctuations $G(\omega) = D_{31}(\omega)$, the Dirichlet kernel. From the results of Section 6.2 it follows that the transforms of $\{z_t\}$ and $\{I_t\}$, $J_z(\omega)$ and $J_I(\omega)$, respectively, are related by

$$J_z(\omega) \cong \{1 - G(\omega)\}J_I(\omega)$$

$$= \{1 - D_{31}(\omega)\}J_I(\omega).$$

Thus the periodogram of $\{z_t\}$ consists approximately of that of $\{I_t\}$ multiplied by $\{1 - D_{31}(\omega)\}^2$, the corresponding power transfer function. A graph of this function appears in Figure 7.2. The function has the value 1 whenever $D_{31}(\omega)$ vanishes, that is, at $\omega = 2\pi k/31$, $k = 1, 2, \ldots$. There are alternate peaks and troughs between these points, occurring roughly at $(2k+1)\pi/31$, $k = 1, 2, \ldots$, with the first peak almost reaching the value 1.5. The corresponding period is $62/3 = 20\frac{2}{3}$ years, and it follows that the periodogram values at around this frequency are considerably amplified. The first trough falls almost to the value 0.75 and occurs near $\omega = 5\pi/31$, which corresponds to a period of $62/5 = 12.4$ years. The periodogram is correspondingly attenuated at approximately this frequency.

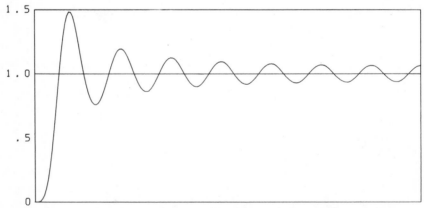

Figure 7.2   The power transfer function $\{1 - D_{31}(\omega)\}^2$.

The enhancement of periods around 21 years and the attenuation of periods around 12 years could easily lead one to infer the existence of a peak in the periodogram at about 21 years. The possibility that spurious periodicities may be introduced into data by operations involving linear filters was raised by Slutsky in 1927 (see, for instance, Slutsky, 1937), and the phenomenon is known by his name. Schuster (1898, Section 13) was also aware that linear filtering may cause such distortions of the transform of a series. Table 7.1 shows the effect on the periodogram of the wheat-price series. The first two columns contain the integer periods from 2 to 36 years and the periodogram ordinates given by Beveridge (1921), ordered by the magnitudes of the ordinates. [Some of these values were revised by Beveridge (1922), but not substantially.] The last three columns show the periods, the periodogram ordinates corrected by dividing by $\{1 - D_{31}(\omega)\}^2$, and the rank of the ordinate before correction. As might be expected, the 20-year period moves down several places, while the 12- and 13-year periods move up. Granger and Hughes (1971) describe the effect of a similar correction and find that the largest peak is moved from a 15.2-year period, close to Beveridge's figure of 15.3 years, to 13.3 years.

A different procedure is suggested by the model $x_t = T_t I_t$. All three terms are strictly positive, and we may write

$$\log x_t = \log T_t + \log I_t.$$

Thus the logarithms of the data should consist of a smooth term, $\log T_t$, with an *added* irregular component, $\log I_t$. It is natural to describe T_t as a typical value of $\{x_t\}$ in the neighborhood of t, and then I_t is a dimension-less quantity close to 1. Hence $\log I_t$ fluctuates about 0. Figure 7.3, which shows the logarithms of the orginal data, is a considerable improvement over the original data (Figure 1.3), in that the fluctuations show no definite tendency either to increase or to decrease in magnitude. This is to be expected from the similarly constant amplitude of the oscillations in the index of fluctuations shown in Figure 7.1 (see Exercise 7.2). The logarithms also show some advantages over the index of fluctuations, however, in that the spikiness of the peaks has been reduced. It was shown in Section 5.5 that nonsinusoidal behavior such as a marked disparity between the natures of the peaks and those of the troughs of a series introduces into its Fourier transform structure that is not revealed by the periodogram. Thus it is preferable when possible to transform a series so as to reduce or remove such disparities. The logarithmic transformation goes some way toward achieving this for the wheat-price series.

Beveridge (1921) argued that the early part of the series is unreliable as it is based on data from relatively few sources, and that the later part of the series is of a different nature because of economic changes in the nineteenth century. He chose the years 1545 to 1844 for his periodogram analysis, giving a series of 300 terms. Figure 7.4 shows the base-10 logarithms of the periodograms of the index of fluctuations and the *natural*

Table 7.1 Beveridge's periodogram for integer periods, before and after correction for the power transfer function

Original Analysis		After Correction		
Period (years)	Periodogram Ordinate	Period (years)	Periodogram Ordinate	Previous Rank
15	47.28	15	50.48	1
11	40.93	11	46.51	2
20	32.44	36	36.92	7
17	29.35	13	36.41	5
13	27.81	12	25.99	9
24	26.48	17	24.59	4
36	26.27	35	22.96	10
16	20.14	20	22.39	3
12	20.11	16	18.90	8
35	17.52	24	18.51	6
18	17.26	34	13.50	15
25	14.95	18	13.23	11
6	12.29	7	12.12	16
8	12.05	6	11.53	13
34	11.04	8	11.31	14
7	10.43	25	10.81	12
30	7.86	30	7.38	17
23	7.54	23	5.15	18
22	7.50	22	5.06	19
21	6.33	10	5.06	21
10	5.39	31	5.04	23
29	5.05	5	4.73	24
31	5.04	29	4.46	22
5	4.43	21	4.28	20
4	3.44	4	3.23	25
26	3.35	33	3.22	27
33	2.82	26	2.52	26
9	2.39	9	2.00	28
28	2.00	28	1.67	29
19	1.50	32	1.12	32
27	1.25	19	1.08	30
32	1.05	27	.99	31
3	.23	3	.25	33
14	.02	14	.02	34
2	.01	2	.01	35

logarithms (see Exercise 7.2) of the original data. The index of fluctuations was first corrected for its typical value 1 and then tapered 20%. The logarithms were detrended by subtracting the least squares straight line and then similarly tapered 20%. Both series were extended by zeros to length 512 and transformed using the fast Fourier transform program listed in the appendix to Chapter 4.

The periodogram of the logarithms is considerably larger than that of

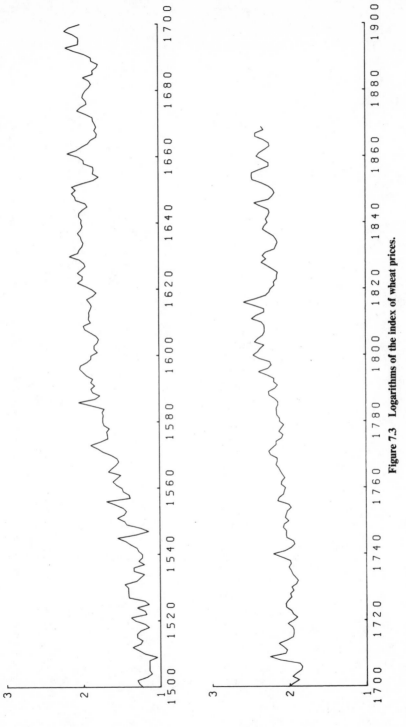

Figure 7.3 Logarithms of the index of wheat prices.

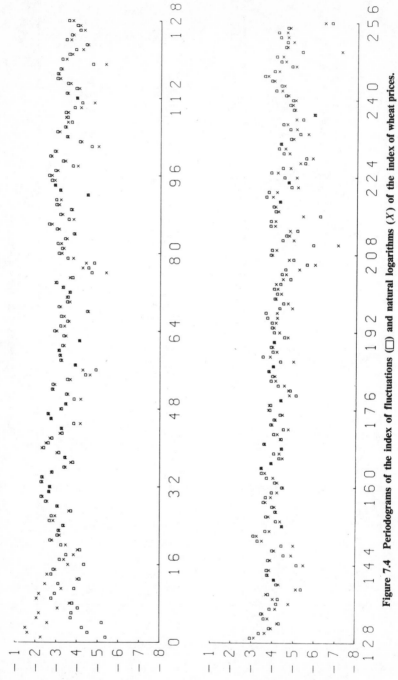

Figure 7.4 Periodograms of the index of fluctuations (□) and natural logarithms (X) of the index of wheat prices.

the index of fluctuations at low frequencies, since even after a trend line has been removed obvious low-frequency terms are still present. From the 17th to the 33rd ordinate the periodogram of the index of fluctuations is generally the larger; this is the first interval in which the transfer function causes enhancement. The ordering tends to be reversed over the next few ordinates, but not consistently. Over the rest of the periodogram there is a general tendency for the periodogram of the index of fluctuations to be the larger, possibly because of the less sinusoidal nature of the oscillations. (The disparity between the values shown in Table 7.1 and those in Figure 7.4 consists of a factor of 10^4, caused by differing definitions of the index of fluctuations.)

Exercise 7.1   Removing a Multiplicative Trend

Suppose that a series $\{x_t\}$ may be represented as $x_t = T_t I_t$, where $\{T_t\}$ is a smooth trend series, and $\{I_t\}$ is an irregular series consisting of small fluctuations around 1. Suppose that T_t is approximated by y_t as in (1). Show that $z_t = x_t / y_t$ satisfies

$$z_t \cong 1 + I_t - \sum_u g_u I_{t-u}.$$

(*Hint*: Since T_t is smooth, we may assume that

$$y_t \cong T_t \sum_u g_u I_{t-u},$$

and hence

$$z_t \cong \frac{I_t}{\sum_u g_u I_{t-u}}.$$

Write $I_t = 1 + \varepsilon_t$, where $|\varepsilon_t| \ll 1$, and expand using a Taylor series.)

Exercise 7.2   (Continuation) The Logarithmic Transformation

Suppose that the series $\{x_t\}$ is as in Exercise 7.1. Let

$$y_t = \sum_u g_u \log_e x_{t-u}$$

and $z_t = \log_e x_t - y_t$. Show that

$$z_t \cong I_t - \sum_u g_u I_{t-u}.$$

Note that this implies that the fluctuations in the natural logarithms of a series should be approximately the same as the quotients defined in Exercise 7.1—for example, Beveridge's index of fluctuations.

7.2 ANALYSIS OF SEGMENTS OF A SERIES

To investigate the consistency of the periodicities in the wheat-price series Beveridge (1921) also gives some terms from the periodograms of the two halves of the series (as suggested by Schuster, 1898), from 1545 to 1694 and from 1695 to 1844, respectively. Corresponding periodograms of the natural logarithms of the series are shown in Figure 7.5. Each half of the transformed series was detrended and tapered as in Section 7.1. The two subseries were then extended to length 512 (rather than 256, which would have been sufficient) before Fourier transformation, so that the transform is evaluated at the same frequencies as for the whole series. Notice, however, that the resulting grid is more than three times finer than that of the Fourier frequencies for the subseries length, 150, and hence the individual periodograms are smoother than might otherwise be expected.

The two periodograms have the same general shape. They are large at low frequencies and show a broad peak from around the 32nd to the 48th ordinates, corresponding to periods between $512/32 = 16$ years and $512/48 = 10\frac{2}{3}$ years. There is then a gentle decline over the rest of the periodogram, with small fluctuations superimposed. However, the fine structures of the two periodograms are quite unrelated: a local peak in one is just as likely to be matched by a local trough as a local peak in the other. We may describe this by saying that the fine structure in the periodograms is not *repeated* from one segment to the next, but that there is a *statistical regularity* or *consistency* in the broad features.

Thus the fine structure of the periodogram of these data is not characteristic of the series as a whole, but depends on the segment used for its computation. On the other hand, it appears that the broad features of the periodogram do not vary in this way and are characteristic of the series as a whole. Thus, if we are interested in the properties of the whole series, we should ignore the fine structure and look only at the broad features of the periodograms. These features of the periodogram of a series that are repeated from one segment to another, that is, the features that show statistical regularity, are known as the *spectrum* of the series. In the present case, we may regard each periodogram as consisting of the same relatively smooth curve with different sets of fluctuations superimposed. This smooth curve is therefore the spectrum of the series, and it is this that we need to extract. As with the complex demodulation problem discussed in Section

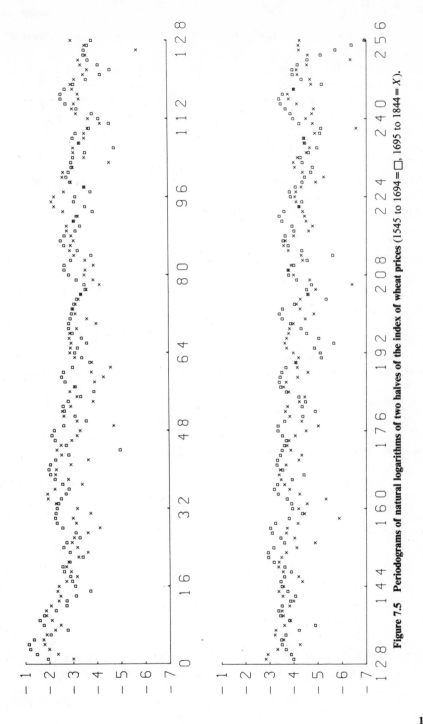

Figure 7.5 Periodograms of natural logarithms of two halves of the index of wheat prices (1545 to 1694 = □, 1695 to 1844 = X).

6.1, this is usually accomplished by *linear filtering,* or *smoothing,* the periodogram.

By contrast, the variable-star series analyzed in Chapters 2 and 5 *does* contain pure, persistent sinusoids. Periodograms computed from different segments of this series would show peaks at precisely the same frequencies. For these data the peaks in the periodogram *are* repeated and *are* characteristic of the series as a whole. The spectrum of this series therefore consists of these peaks. If the periodograms of successively longer stretches of data from this series were computed, the peaks would become narrower, and we therefore regard the spectrum of a series such as this as consisting of spikes of infinitesimal width at the frequencies that are present. This idea is made more precise in Chapter 8.

7.3   SMOOTHING THE PERIODOGRAM

If we wish to see the common features of the two periodograms of Figure 7.5, a simple procedure would be to graph the average of the two ordinates at each frequency. We could either average the logarithms or plot the logarithm of the average of the original periodogram ordinates. The latter is preferable, since averaging the logarithms puts more weight on small values than on large ones, and the small values are most subject to perturbations of all kinds, particularly leakage from other frequencies. More generally, we could divide the data into a number of segments, calculate a periodogram for each, and average them. This procedure was first suggested by Bartlett (1948; see also Bartlett, 1950, and Kendall, 1948). To see more clearly what this procedure does, we need to derive an alternative representation of the periodogram.

For convenience we use the complex representation of the periodogram. To simplify the expressions below, we shall assume that x_t is the *difference* between the tth observation and the series mean. This is equivalent to assuming that $\bar{x} = n^{-1}(x_0 + \cdots + x_{n-1}) = 0$. The periodogram is

$$I(\omega) = \frac{1}{2\pi n} \left| \sum_{t=0}^{n-1} x_t \exp(-it\omega) \right|^2$$

$$= \frac{1}{2\pi n} \sum_t \sum_u x_t x_u \exp\{-i(t-u)\omega\},$$

since the squared modulus of a complex number is the number multiplied by its complex conjugate. The periodogram is thus itself a Fourier series, in which the coefficient of $\exp(-ir\omega)$ is the sum of $x_t x_u$ over all pairs (t, u) for which $t - u = r$. A little manipulation yields

$$I(\omega) = \frac{1}{2\pi} \sum_{|r|<n} c_r \exp(-ir\omega), \qquad (2)$$

where

$$c_r = \begin{cases} \dfrac{1}{n} \sum_{t=r}^{n-1} x_t x_{t-r}, & r \geqslant 0, \\ \\ c_{-r}, & r < 0. \end{cases}$$

(See Exercise 7.3.) The quantity c_r is the *autocovariance* of $\{x_t\}$ at lag r. Because of the symmetry of $\{c_r\}$, (2) may also be written as

$$I(\omega) = \frac{1}{2\pi} \left(c_0 + 2 \sum_{r=1}^{n-1} c_r \cos r\omega \right). \qquad (3)$$

Suppose that we divide the data into k segments, each of length m, where $n = km$. Let $I_j(\omega)$ be the periodogram of the jth segment, and $\{c_{j,r}\}$ the corresponding autocovariances. The average is

$$g(\omega) = \frac{1}{k} \sum_{j=1}^{k} I_j(\omega)$$

$$= \frac{1}{2\pi} \sum_{|r|<m} \left(k^{-1} \sum_{j=1}^{k} c_{j,r} \right) \exp(-ir\omega)$$

$$= \frac{1}{2\pi} \sum_{|r|<m} \left(1 - \frac{|r|}{m} \right) \frac{1}{k(m-|r|)} \sum_j m c_{j,r} \exp(-ir\omega). \qquad (4)$$

Now $\sum_j m c_{j,r}$ is, like nc_r, a sum of products of the form $x_t x_{t+r}$, but not all such products, for the term $x_t x_{t+r}$ is included only if x_t and x_{t+r} fall in the same segment of the series. There are $m - |r|$ terms in each $c_{j,r}$, and thus

$$\frac{1}{k(m-|r|)} \sum_j m c_{j,r}$$

is the average of these products. It seems rational to replace this part of (4) by $nc_r / (n - |r|) = c_r / (1 - |r|/n)$, the average of *all* available products of the form $x_t x_{t+r}$. This leads to the modified function

$$g_B(\omega) = \frac{1}{2\pi} \sum_{|r|<m} \frac{1-|r|/m}{1-|r|/n} c_r \exp(-ir\omega) \tag{5}$$

$$= \frac{1}{2\pi} \sum_{|r|<m} w_r c_r \exp(-ir\omega), \tag{6}$$

say, where

$$w_r = \frac{1-|r|/m}{1-|r|/n}.$$

The function $g_B(\omega)$ defined in (5) is known as the *Bartlett spectrum estimate*, and the numbers $\{w_r\}$ are the corresponding *lag weights*.

   This differs from the periodogram in that all terms for $|r| \geqslant m$ have been omitted. The remaining terms are progressively reduced in magnitude, and it is clear that the resulting function is smoother than the periodogram. Furthermore, by varying the *truncation point m* (which need no longer be required to divide n) we may vary the smoothness of the function.

   This modification of the periodogram to make it smoother is analogous to the application of *convergence factors* to a truncated Fourier series (see Section 6.4), and also to the use of a *data window* (see Section 5.2) on a data series. In each case a series of numbers is tapered at its end to make its Fourier transform smoother. It is due partly to an historical accident and partly to a difference in interpretation that the classes of tapers used are different in each procedure.

   As in the other two cases, the significant property of the lag weights is that they decay smoothly from the value 1 at $r=0$ to 0 at $r = \pm m$, and many such sets of weights have been used to construct spectrum estimates. A list of the most commonly used sets of weights and their properties is given by Anderson (1971, Chapter 9). Note that, in contrast with the tapers used as data windows and as convergence factors, the Bartlett weights begin their decay linearly since, for small r, $\omega_r \cong 1 - |r|(m^{-1} - n^{-1})$. All the other sets of weights used to construct spectrum estimates have the property that the weights stay closer to 1 for small r, and typically behave like $1 - \alpha r^2$ for some $\alpha > 0$. For this reason the Bartlett estimate is rarely used in practice.

   The most obvious way to compute a spectrum estimate is by direct application of (3) and (5) (see, for instance, Blackman and Tukey, 1959). As we show in the next section, the autocovariances $\{c_r\}$ may in fact be computed efficiently using the fast Fourier transform and its inverse. The transform (5) may then be computed by a third application of the fast

Fourier transform. This method involves less computation if the truncation point m is sufficiently large. A simpler way to construct spectrum estimates involving even less computation is described in Section 7.5.

Exercise 7.3   Alternative Expression for the Periodogram

Verify that the periodogram may also be written as in (2), where the autocovariances $\{c_r\}$ are defined by (3).

7.4   COMPUTING AUTOCOVARIANCES AND LAG-WEIGHTS SPECTRUM ESTIMATES

To compute the rth autocovariance of $\{x_t\}$, defined in (3), directly from that formula requires $n-r$ multiplications and $n-r-1$ additions. Computation of $c_0, c_1, \ldots, c_{m-1}$ therefore requires $m\{n-\frac{1}{2}(m-1)\}$ multiplications and $m\{n-\frac{1}{2}(m+1)\}$ additions. However, we see from (2) that the periodogram is (proportional to) the Fourier transform of $\{c_r\}$, and this raises the possibility of using the fast Fourier transform to obtain the auto-covariances (see, for instance, Gentleman and Sande, 1966).

If we evaluate the periodogram at the Fourier frequencies $\omega_j = 2\pi j/n$, we find that

$$I(\omega_j) = \frac{1}{2\pi} \sum_{r=0}^{n-1} (c_r + c_{n-r}) \exp(-ir\omega_j) \tag{7}$$

(see Exercise 7.4), provided we adopt the convention that $c_r = 0$ for $|r| \geqslant n$. Thus

$$c_r + c_{n-r} = \frac{2\pi}{n} \sum_{j=0}^{n-1} I(\omega_j) \exp(ir\omega_j),$$

a computation that may be carried out using the fast Fourier transform. The zero-lag autocovariance c_0 is obtained from this analysis, since $c_n = 0$. For small r, c_{n-r} is small, since it is a sum of r terms divided by n. Thus $c_r + c_{n-r} \cong c_r$, and the analysis also gives approximations to the auto-covariances at small lags.

Often, however, the periodogram is evaluated at the frequencies $\omega_j' = 2\pi j/n'$, the Fourier frequencies for a series of length $n' > n$ (see Section 4.5). At these frequencies we find that

$$I(\omega_j') = \frac{1}{2\pi} \sum_{r=0}^{n-1} (c_r + c_{n'-r}) \exp(-ir\omega_j'). \tag{8}$$

Inversion now gives

$$c_r + c_{n'-r} = \frac{2\pi}{n'} \sum_{j=0}^{n'-1} I(\omega_j') \exp(ir\omega_j'),$$

and thus $c_0, c_1, \ldots, c_{n'-n}$ are obtained exactly.

The computational cost of each fast Fourier transform of a series of length n' is approximately $2n' \log_2 n'$ multiplications and $3n' \log_2 n'$ additions (using the radix-2 algorithm; for exact figures for algorithms of radix 2, 8, and 16 see Bergland, 1968). Computation and inversion of the periodogram therefore requires approximately $4n' \log_2 n'$ multiplications and $6n' \log_2 n'$ additions. Thus the fast Fourier transform approach leads to computational savings only if m is somewhat larger than $4 \log_2 n'$. For instance, for the wheat-price series of length $n = 300$ we need to use $n' = 512$, and in this case the break-even point is around $m = 70$. Since these counts of arithmetic operations ignore the cost of incrementing indices and similar overhead items, the true break-even point is probably rather higher.

If we proceed to compute a spectrum estimate as in (6) for some given set of weights $\{w_r\}$, the computations may again be carried out using the fast Fourier transform. Suppose that we wish to compute the spectrum estimate at the frequencies $\omega_j' = 2\pi j / n'$, $j = 0, \ldots, n'/2$. We no longer assume that $n' > n$. In fact, since a spectrum estimate is a smooth function, it is often sufficient to evaluate it at relatively few frequencies, and $n' = m$ will frequently be enough. The direct calculation based on the real form of (6),

$$\frac{1}{2\pi}\left(c_0 + 2 \sum_{r=1}^{m-1} w_r c_r \cos r\omega \right)$$

requires around $mn'/2$ multiplications and additions. We ignore the cost of computing the required cosines, since computational considerations are usually important only in the case of repeated calculations, and the values may be tabulated to avoid recomputation. To use the fast Fourier transform we have to extend the weighted series of autocovariances to length n'. We let

$$d_r = \begin{cases} w_r c_r, & 0 \leqslant r < m \\ 0, & m \leqslant r < n' \end{cases}$$

and $e_r = d_r + d_{n'-r}$, $r = 0, 1, \ldots, n'-1$. An argument similar to that used to obtain (7) shows that the transform of $\{e_r\}$, suitably rescaled, is the required spectrum estimate. The computational cost is roughly $2n' \log_2 n'$ multiplications and $3n' \log_2 n'$ additions, and this is less than $mn'/2$ if $m > 4 \log_2 n'$. Thus the fast Fourier transform approach leads to computational savings only if m, the truncation point, is relatively large and n', the number of frequencies at which the estimate is to be calculated (strictly, twice this number), is relatively small.

The computation of autocovariances is one of the tasks mentioned in Section 4.4 in which a transform calculated by the fast Fourier transform algorithm does not need to be unscrambled. The sequence of operations is as follows:

(i)   transform $x_0, \ldots, x_{n-1}, 0, \ldots, 0$ to obtain $J(\omega_j'), j = 0, \ldots, n' - 1$,

(ii)   convert to $|J(\omega_j')|^2$,

(iii)   invert $|J(\omega_j')|^2, j = 0, \ldots, n' - 1$, to obtain $c_0, \ldots, c_{n'-n}$.

Step (i) may be carried out using the Sande-Tukey algorithm *without* unscrambling, since step (ii) can be carried out just as well on the scrambled as on unscrambled transform. The inversion in step (iii) may then be computed efficiently using the twiddled Cooley-Tukey algorithm, since the values to be inverted,

$$|J(\omega_j')|^2, \qquad j = 0, \ldots, n' - 1,$$

are already in scrambled form. Notice, however, that if a lag-weights spectrum estimate is then computed from the autocovariances by a third application of the fast Fourier transform, unscrambling is needed if the result is to be obtained in the correct order.

Exercise 7.4   *Inverting the Periodogram*

Show that the periodogram, evaluated at a Fourier frequency, may be written as in (7). [*Hint*: Use (2), the symmetry of $\{c_r\}$, and the special properties of the Fourier frequencies.]

7.5 ALTERNATIVE REPRESENTATIONS OF A SPECTRUM ESTIMATE

Suppose that a spectrum estimate $g(\omega)$ is given by

$$g(\omega) = \frac{1}{2\pi} \sum_{|r| < n} w_r c_r \exp(-ir\omega), \tag{9}$$

and that none of the lag weights $\{w_r : -n < r < n\}$ is assumed to vanish. By the integral inversion formula (see Exercise 3.5) applied to the periodogram,

$$c_r = \int_{-\pi}^{\pi} I(\lambda) \exp(ir\lambda) \, d\lambda.$$

Then

$$g(\omega) = \int_{-\pi}^{\pi} W_n(\omega - \lambda) I(\lambda) \, d\lambda, \tag{10}$$

where

$$W_n(\lambda) = \frac{1}{2\pi} \sum_{|r| < n} w_r \exp(-ir\lambda)$$

(see Exercise 7.5). Hence any spectrum estimate of form 9 may be written as an integral average of the periodogram. Equation 10 is the integral analog of the (discrete) linear filters described in Section 6.2.

The function $W_n(\lambda)$ is called the *spectral window* of the spectrum estimate. The spectral window of the Bartlett estimate cannot be written in closed form, but for a modified version with weights

$$w_r = \begin{cases} 1 - |r|/m, & |r| < m \\ 0, & |r| \geqslant m \end{cases}$$

the spectral window is

$$W_n(\lambda) = \frac{1}{2\pi m} \sum_{|r| < m} (m - |r|) \exp(-ir\lambda)$$

$$= \frac{1}{2\pi m} \left| \sum_{r=0}^{m-1} \exp(-ir\lambda) \right|^2$$

$$= \frac{m}{2\pi} D_m(\lambda)^2,$$

where $D_m(\lambda)$ is the Dirichlet kernel (see Section 3.4). The central peak is of height $m/2\pi$, and the first zeros on either side are at $\lambda = \pm 2\pi/m$. However, a sizable proportion of the mass of $W(\lambda)$ is contained not in the main peak but in the sidelobes. These decay slowly, and hence the spectrum estimate at frequency ω may be influenced by periodogram values at some distance from ω. We have seen that periodogram ordinates at different frequencies often differ by several orders of magnitude, and hence the estimated spectrum in a frequency band where there is little power can be swamped by such *leakage* from a band with high power, even if these bands are not adjacent.

This source of leakage is quite distinct from the ones mentioned in preceding chapters. It was shown in Chapter 5 that if no data window is used there may be substantial leakage in the periodogram itself. However, even if leakage in the periodogram is controlled by the use of a data window, leakage may appear in a smoothed version of the periodogram because of sidelobes in the special window.

The sidelobes of the modified Bartlett window are larger and decay more slowly than those of the most commonly used estimates (see Anderson, 1972, Chapter 9), and for this reason it is rarely used. However, such sidelobes are bound to exist for any spectrum estimate for which $w_r = 0$ for $|r| \geqslant m$ (that is, with a *truncation point m*), for $W(\lambda)\exp\{i(m-1)\lambda\}$ is a polynomial of degree $2m - 1$ in $\exp(-i\lambda)$ and hence may vanish at only $2(m-1)$ frequencies. For a given m the best we can do is to place these zeros so that $W(\lambda)$ remains small for λ not close to 0. It is desirable for $W(\lambda)$ to be nonnegative, since otherwise negative values of $g(\omega)$ may arise. This requires that the zeros of the polynomial occur in pairs.

Equation 10 suggests a different way of constructing spectrum estimates. If we choose any suitable function $W(\lambda)$ and let

$$g(\omega) = \int_{-\pi}^{\pi} W(\lambda) I(\omega - \lambda) d\lambda, \tag{11}$$

the result is a smoothed version of the periodogram. Furthermore, it may be written in form (9) with

$$w_r = \int_{-\pi}^{\pi} W(\lambda) \exp(ir\lambda) d\lambda \tag{12}$$

(see Exercise 7.5). This implies that $g(\omega)$ is also given by (10), but notice that $W(\lambda)$ and $W_n(\lambda)$ need not be the same, for $\{w_r\}$ are the Fourier coefficients of $W(\lambda)$, and thus $W_n(\lambda)$ is the nth partial sum of the Fourier series of $W(\lambda)$.

Daniell (1946) suggested an estimate of form (11), with

$$W(\lambda) = \begin{cases} \dfrac{1}{2\delta}, & |\lambda| < \delta, \\ 0, & |\lambda| \geqslant \delta. \end{cases}$$

The resulting estimate $g_D(\omega)$ is simply the integral average of the periodogram over an interval of length 2δ surrounding ω. This is the integral analog of the simple moving average filters discussed in Section 6.3 and may, in the same way, be applied successively to build up more complex filters, as we show in Section 7.6.

The estimate (11) cannot be computed directly from the definition (11). It may be computed as a lag-weights estimate through (12), or the integral (11) could be approximated by a numerical quadrature formula. If $W_n(\lambda)$ is available, $g(\omega)$ may also be calculated as

$$g(\omega) = \frac{2\pi}{n'} \sum_u W_n(\omega - \omega_u') I(\omega_u'), \qquad (13)$$

where $\omega_u' = 2\pi u/n'$ are the Fourier frequencies for any $n' \geqslant 2n - 1$. Equation 13 may be regarded as an exact quadrature formula for (10) or (11). The result of Exercise 7.6 shows that, if the estimate is in fact a lag-weights estimate with truncation point m, then $n' \geqslant n + m - 1$ is sufficient for (13) to hold.

Equation 13 shows that the estimate $g(\omega)$ may be obtained from the periodogram, evaluated on a sufficiently fine grid, by a *linear filtering* operation (see Section 6.2). This is the simplest way to describe such an estimate and is also a useful computational route, for we may use the fast Fourier transform to compute the required periodogram ordinates and then apply the linear filter directly. If the filter has weights $\{g_u\}$, which in this context we shall call *spectral weights*, the formula is

$$g(\omega) = \sum_u g_u I(\omega - \omega_u').$$

We term such an estimate a *discrete spectral average*. It is easily seen that this is also a lag-weights estimate with lag weights $w_r = G(-\omega_r)$, where $G(\omega)$ is the *transfer function* of the filter $\{g_u\}$,

$$G(\omega) = \sum g_u \exp(-iu\omega).$$

We may now choose the spectral weights directly so as to make the filtering operation have the required smoothing effect and yet be computationally simple. The choice of a set of spectral weights is discussed in the next section.

Exercise 7.5 *Alternative Representations*

(i) Verify (10). Note that $W_n(\lambda)$ is a periodic function of λ with period 2π.

(ii) Verify that $g(\omega)$ defined by (11) satisfies (9) with w_r given by (12).

Exercise 7.6 Discrete Spectral Averages

Suppose that $g(\omega)$ is defined by (10), and that $\omega_r = 0$ for $|r| \geqslant m$. Note that $m \leqslant n$. Show that (13) holds for any $n' \geqslant n + m - 1$. [*Hint:* Substitute the definition of $W_n(\lambda)$ below (10) in (13), and use the orthogonality relations to show that the result simplifies to (9).]

Note that (13) does *not* hold in general if $n' < n + m - 1$.

Exercise 7.7 Spectrum Estimation by Complex Demodulation

Suppose that a series $\{x_0, \ldots, x_{n-1}\}$ is demodulated at frequency ω and then filtered using weights $\{g_r\}$. If $R_t(\omega)$ is the instantaneous amplitude at time t, show that

$$g(\omega) = \frac{\sum\limits_{t=0}^{n-1} R_t(\omega)^2}{8\pi n \sum\limits_u g_u^2}$$

is approximately the spectrum estimate

$$\frac{1}{2\pi} \sum_{|r|<u} w_r c_r \exp(-ir\omega),$$

where

$$w_r = \frac{\sum\limits_u g_u g_{u-r}}{\sum\limits_u g_u^2}.$$

Show that the corresponding spectral window is

$$W_n(\lambda) = \frac{|G(\lambda)|^2}{2\pi \sum\limits_u g_u^2},$$

where $G(\omega)$ is the transfer function of the filter $\{g_u\}$.

Note that, for any lag-weights estimate with trucation point m, if $W_n(\lambda)$ is nonnegative it may be factorized in this way so that the span of the filter is m. Thus any such estimate may be interpreted as an approximation to an estmate of this kind. Note also that this procedure may be regarded as a generalization of the original Bartlett procedure. For more details see Bingham, Godfrey, and Tukey (1967).

7.6 CHOICE OF A SPECTRAL WINDOW

Four basic factors need to be taken into account when choosing a spectral window:

(i) resolution, or bandwidth,
(ii) stability,
(iii) leakage, and
(iv) smoothness.

Resolution is the ability of a spectrum estimate to represent fine structure in the frequency properties of the data, such as narrow peaks in the spectrum. Because of the averaging involved in computing a spectrum estimate, a spike or narrow peak in the periodogram is spread out into a broader peak. This peak is roughly an image of the spectral window of the estimate, and its "width," suitably defined, is the *bandwidth* of the estimate (see Section 8.6). If the spectrum of a series contains two narrow peaks closer together than the bandwidth of the estimate used, the resulting broad peaks in the estimated spectrum overlap and form a single peak. In this case we say that the estimate fails to *resolve* these peaks.

The *stability* of a spectrum estimate is the extent to which estimates computed from different segments of a series agree or, in other words, the extent to which irrelevant fine structure in the periodogram is eliminated. Resolution and stability are conflicting requirements, because high stability requires averaging over many periodograms ordinates, whereas this reduces resolution. A statistical treatment of the stability of spectrum estimates is given in Section 8.5.

Leakage has been discussed in the context of the Bartlett estimate in Section 7.5. It is caused by sidelobes in the spectral window, which are always present in a lag-weights estimate with a nontrivial truncation point (that is, $m < n$). However, if we use the computationally simpler discrete spectral average estimates described in Section 7.5, we may avoid leakage problems entirely.

The *smoothness* of a spectrum estimate is a less tangible property that may add a further conflict. For instance, suppose that we define the bandwidth of a window to be the length of the interval on which it is nonzero. Then it may be shown, using the results of the next chapter, that the most stable estimate for a given bandwidth is the corresponding Daniell estimate. However, if the periodogram contains one or more large spikes, as is often the case, these will be flattened out into rectangles (or *boxcars*). The result can hardly be called smooth, and the angular appearance of the spectrum estimate makes visual assessment more difficult. However, successive applications of the Daniell procedure can yield a satisfactory estimate, as we now show.

7.7   EXAMPLES OF SMOOTHING THE PERIODOGRAM

Figure 7.6 shows the result of smoothing the periodogram of the logarithms of the wheat-price series, plotted as usual on a logarithmic scale. The periodogram itself is shown in Figure 7.4, and its computation is described in Section 7.1. The frequencies at which it was calculated are $\omega'_j = 2\pi j / n' = 2\pi j / 512, j = 0, \ldots, 256$. The smoothing was carried out using a modified discrete Daniell procedure,

$$g_1(\omega) = \tfrac{1}{8}\left\{ \tfrac{1}{2}I(\omega + \omega'_4) + \sum_{u=-3}^{3} I(\omega - \omega'_u) + \tfrac{1}{2}I(\omega - \omega'_4) \right\}, \qquad (14)$$

which is the simple moving average of length 8 (see Section 6.3). The spectral weights are shown inset in Figure 7.6. Since the curve is smoother than the orginal periodogram, it may be graphed with a smaller scale on the frequency axis. At the same time the smoothing replaces the many small values of the periodogram by averages of their neighbors, and hence the range of orders of magnitude of the ordinates (their *dynamic range*) is reduced. Thus the vertical scale may be increased to show the nature of the peaks and troughs more clearly. Note that once the periodogram has been smoothed in this way, it can no longer be interpreted as the squared amplitude of a sinusoidal component in the data. Accordingly, the curve in Figure 7.6 is the result of smoothing the periodogram $I(\omega)$ itself, and not

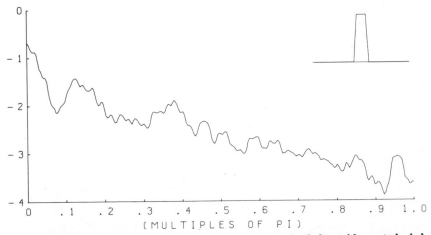

Figure 7.6   Smoothed periodogram of logarithms of wheat-price index, with spectral window inset. (Modified Daniell filter.)

the squared amplitude function

$$\tilde{R}(\omega)^2 = \frac{8\pi}{n} I(\omega),$$

which we have previously used.

The graph in Figure 7.6 is still by no means smooth, and in a number of places we see an image of the spectral window, where a relatively large periodogram ordinate has not been smoothed out. Figure 7.7 shows the result of applying a second modified Daniell filter to Figure 7.6, this time a simple moving average of length 16,

$$g_2(\omega) = \tfrac{1}{16} \left\{ \tfrac{1}{2} g_1(\omega + \omega'_8) + \sum_{u=-7}^{7} g_1(\omega - \omega'_u) + \tfrac{1}{2} g_1(\omega - \omega'_8) \right\}.$$

As was shown in Section 6.2, the result is equivalent to applying a combined filter directly to the periodogram, and the corresponding spectral weights are shown inset. The many local fluctuations in Figure 7.6 have been smoothed out, but the broad features remain. The two peaks in the spectrum at frequencies around 0.14π and 0.37π are quite clear. It appears that there is no other structure to the spectrum, apart from the decrease in spectral power at higher frequencies.

In Figure 7.7 the trough between the peak at zero frequency and the peak at frequency 0.14π is not as well defined as it is in Figure 7.6. This is

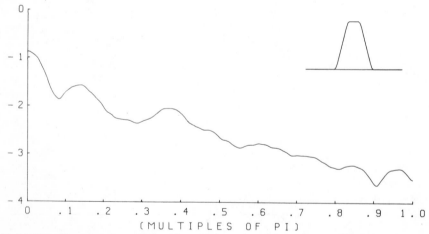

Figure 7.7 Smoothed periodogram of logarithms of wheat-price index. (Two applications of modified Daniell filters.)

because the bandwidth is now large enough for the two peaks to begin to overlap, with a corresponding slight loss of resolution. This could be avoided by *prewhitening* the data (Blackman and Tukey, 1959). Prewhitening is a technique for reducing the dynamic range of a spectrum by filtering the data before estimating the spectrum. It reduces the problem of leakage and allows the use of a more stable estimate with lower resolution. In the present case the object of prewhitening would be to eliminate the peak at zero frequency and the decrease in power at higher frequencies. Replacing the data by their first differences $y_t = x_{t+1} - x_t$ would be the simplest solution. This is a linear filter with transfer function $G(\omega) = \exp(i\omega) - 1$ and power transfer function $|G(\omega)|^2 = |\exp(i\omega) - 1|^2 = 4(\sin \frac{1}{2}\omega)^2$, which has the right general properties. Computing an analog of Beveridge's index of fluctuations, by subtracting a centered average of 31 terms from each observation, would be an alternative. (Recall that the data being analyzed are the logarithms of the original price index.) However, it was shown in Section 7.1 that this introduces serious distortions into the periodogram and hence into any spectrum estimate based on it. It is essential that the transfer function of any filter used to prewhiten a series be smooth.

It should be noted that in this example prewhitening would help only by removing the need to resolve the peaks at frequencies 0 and 0.14π. In some cases prewhitening is also needed to avoid leakage problems. However, by tapering the data before transforming them and by using a discrete spectral average with no sidelobes, we have avoided the two possible sources of serious leakage.

Figure 7.8 shows estimates of the spectra of the yearly sunspot numbers (Figure 1.2) and the square roots of these numbers (Figure 5.13). They were obtained by smoothing the periodograms of the two series (Figures 5.10 and 5.14, respectively), using three applications of the modified Daniell procedure (14). (The spectrum of the square roots was rescaled so that the peaks are at the same value.)

The spectral weights are again shown inset. They have the same *span* as those in Figure 7.7 (that is, they cover the same number of periodogram ordinates), but thay are smoother and have a narrower peak. The spectrum estimates correspondingly show more rounded but slightly larger fluctuations. Notice that the sunspot series is of length 261, while the wheat-price index is of length 300. Thus, even though the spans of the two sets of spectral weights are the same, fewer Fourier frequencies are covered in the case of the sunspot numbers. The spacing of the Fourier frequencies for the (unextended) series is the bandwidth of the periodogram itself, and thus the stability of the smoothed periodogram depends on the number of Fourier frequencies for the original series length that are covered by the

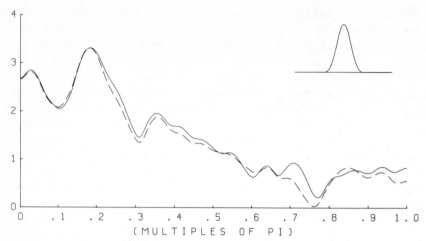

Figure 7.8 **Smoothed periodograms of yearly sunspot numbers and their square roots. (Three applications of a modified Daniell filter.)**

spectral weights. This argument is made more precise in Chapter 8.

The presence of the second harmonic of the basic sunspot cycle is clear, though no higher harmonic is visible. It is also clear that the square root transformation makes the main peak slightly more prominent. The width of the peak has been reduced, and the power at the second harmonic is also less. These modifications indicate that the transformation sought in Section 5.7 has had the required effect of making the main peak more distinct, but not to any great extent. Recall that the change is not visible in the graphs of the periodograms themselves (Figure 5.10 and 5.14); this shows the value of smoothing the periodogram.

Further discussion of the choice of spectral window or lag weights for spectrum estimates may be found in Jenkins (1961) and Parzen (1961) (see also Section 8.6).

Exercise 7.8 Combinations of Filters

Find the weights in the combined filters that were used to obtain the spectrum estimates shown in Figures 7.7 and 7.8. Note that the first set is trapezoidal with some modifications in the angles, and that the second set is quadratic in three sections, again with some modifications at the boundaries. (*Hint*: The moving averages of lengths 8 and 16 can be written as moving averages of lengths 7 and 15, respectively, each followed by a moving average of length 2. Since filtering operations may be applied in order, the averages of length 2 may be applied last.)

APPENDIX

The first program below was used to compute the spectrum estimates displayed in this chapter. The periodogram to be smoothed is read in by subprogram DATIN, after being computed by a program such as that given in the Appendix to Chapter 5. The length of the original series, NOBS, is read in so that the periodogram may be rescaled. (It is assumed that the input periodogram has been calculated as a squared amplitude, rather than the true periodogram.)

The second program was used to compute the spectral weights implicit in the smoothing used by the first program. It uses the same input record (although read in a different format) to ensure that the smoothing procedure used is the same.

Note: These programs also use subprograms DATIN (presented in the Appendix to Chapter 2) and DATOUT (in the Appendix to Chapter 5).

```
C
C     THIS PROGRAM CARRIES OUT PERIODOGRAM SMOOTHING TO
C     OBTAIN A SPECTRUM ESTIMATE.  THE SPECTRUM ESTIMATE IS
C     COMPUTED BY REPEATED SMOOTHING WITH MODIFIED DANIELL
C     WEIGHTS.   THE PROGRAM IS CONTROLLED BY THE FOLLOWING
C     VARIABLES,   WHICH ARE INPUT TO THE PROGRAM  -
C
C     NK    THE NUMBER OF SMOOTHING PASSES
C
C     K     AN ARRAY OF THE SMOOTHING PARAMETERS
C
C     NOTE - THE PERIODOGRAM TO BE SMOOTHED IS READ IN BY
C     SUBROUTINE DATIN.
C
      DIMENSION X(513),Y(513),K(10)
      DATA PI /3.141593/
      CALL DATIN (X,NPGM,START,STEP,7)
      READ(5,1) NOBS,NP2,NK,(K(I),I=1,NK)
    1 FORMAT (10X,13I5)
      CON=FLOAT(NOBS)/(8.0*PI)
      DO 10 I=1,NPGM
   10 X(I)=X(I)*CON
      WRITE(6,6) NK
    6 FORMAT(≠0NUMBER OF MODIFIED DANIELL PASSES IS≠,I5/
     +         ≠ VALUES OF K ARE -≠)
      WRITE(6,1) (K(I),I=1,NK)
      DO 50 I=1,NK
   50 CALL MODDAN (X,Y,NPGM,K(I),1.0)
      START=0.0
      STEP=2.0*PI/FLOAT(NP2)
      CALL DATOUT (X,NPGM,START,STEP,8)
      STOP
      END
```

```
      SUBROUTINE MODDAN (X,Y,N,K,SYM)
C
C    THIS SUBROUTINE APPLIES MODIFIED DANIELL SMOOTHING
C    TO THE SERIES  X.    IT IS USED IN COMPUTING SPECTRUM
C    ESTIMATES FROM PERIODOGRAMS.    IT IS ASSUMED THAT THE
C    INPUT SERIES IS SYMMETRIC ABOUT BOTH OF ITS END VALUES,
C    TO PROVIDE END VALUES OF THE OUTPUT.    PARAMETERS ARE
C
C    NAME    TYPE                         VALUE
C                        ON ENTRY              ON RETURN
C
C    X      REAL ARRAY THE SERIES TO BE     THE SMOOTHED SERIES
C                      SMOOTHED
C
C    Y      REAL ARRAY A SCRATCH VECTOR     THE INPUT SERIES
C
C    N      INTEGER    THE SERIES LENGTH    UNCHANGED
C
C    K      INTEGER    THE HALF-LENGTH      UNCHANGED
C                      OF THE AVERAGE
C
C    SYM    REAL       PLUS OR MINUS ONE    UNCHANGED
C                      ACCORDING TO THE
C                      SYMMETRY OF X
C
C    NOTE - IF  K  IS NOT POSITIVE,  THE INPUT SERIES IS
C    RETURNED UNCHANGED.
C
      DIMENSION X(N),Y(N)
      DO 10 I=1,N
  10  Y(I)=X(I)
      IF (K .LE. 0) RETURN
      LIM=K-1
      CON=1.0/FLOAT(2*K)
      DO 20 I=1,N
      X(I)=Y(I)
      IF (LIM .EQ. 0) GO TO 20
      DO 30 J=1,LIM
  30  X(I)= X(I)+ EXTEND(Y,I-J,N,SYM)
     +           + EXTEND(Y,I+J,N,SYM)
  20  X(I)=(X(I)+(EXTEND(Y,I-K,N,SYM)
     +           + EXTEND(Y,I+K,N,SYM))*0.5)*CON
      RETURN
      END
```

178

```
      FUNCTION EXTEND (X,I,N,SYM)
C
C     THIS FUNCTION RETURNS THE I≠TH TERM IN THE SERIES  X,
C     EXTENDED IF NECESSARY WITH EVEN OR ODD SYMMETRY,
C     ACCORDING TO THE SIGN OF  SYM,  WHICH SHOULD BE EITHER
C     PLUS OR MINUS 1    (THE VALUE ZERO WILL RESULT IN
C     THE EXTENDED VALUE BEING ZERO.)
C
      DIMENSION X(N)
      IF (N .GT. 1) GO TO 10
      WRITE(6,1) N
1     FORMAT(≠0ERROR   -   VALUE OF  N  IN  EXTEND  IS≠,I10)
      STOP
10    J=I
      CON=1
20    IF (J .GE. 1) GO TO 30
      J=2-J
      CON=CON*SYM
30    IF (J .LE. N) GO TO 40
      J=2*N-J
      CON=CON*SYM
      GO TO 20
40    EXTEND=X(J)*CON
      RETURN
      END
```

```
C
C     THIS PROGRAM COMPUTES THE IMPULSE RESPONSE FUNCTION
C     CORRESPONDING TO A NUMBER OF APPLICATIONS OF
C     SUBROUTINE  MODDAN.  THE PROGRAM IS CONTROLLED BY THE
C     FOLLOWING VARIABLES.
C
C  NK      THE NUMBER OF SMOOTHING PASSES
C
C  K       AN ARRAY OF THE SMOOTHING PARAMETERS
C
       DIMENSION WINDOW(101),Y(101),K(10)
       DATA PI /3.141593/
       READ(5,1) NP2,NK,(K(I),I=1,NK)
    1  FORMAT(15X,12I5)
       WRITE(6,6) NK
    6  FORMAT(≠0NUMBER OF MODIFIED DANIELL PASSES IS≠,I5/
      +        ≠ VALUES OF K ARE -≠)
       WRITE(6,1) (K(I),I=1,NK)
       INDEX=2
       DO 70 I=1,NK
   70  INDEX=INDEX+K(I)
       LIM=2*INDEX-1
       DO 80 I=1,LIM
   80  WINDOW(I)=0.0
       WINDOW(INDEX)=1.0
       DO 90 I=1,NK
   90  CALL MODDAN (WINDOW,Y,LIM,K(I),1.0)
       STEP=2.0*PI/FLOAT(NP2)
       START=-STEP*FLOAT(INDEX-1)
       CALL DATOUT (WINDOW,LIM,START,STEP,7)
       G=0.0
       DO 100 I=1,LIM
  100  G=G+WINDOW(I)**2
       WRITE(6,2) G
    2  FORMAT(≠0SUM OF SQUARES OF WEIGHTS IS≠,E15.7)
       STOP
       END
```

8

SOME STATIONARY
TIME SERIES THEORY

In Chapter 7 the spectrum of a time series was defined to be the aspects of its periodogram that show statistical regularity and are characteristic of the series as a whole. This concept is statistical in nature, and in this chapter we discuss a class of theoretical models that show such statistical regularity, namely, the class of *stationary time series*. For a stationary time series we may define a *theoretical spectrum* and describe the extent to which the spectrum estimates described in Chapter 7 give approximations to it. Some properties of the sampling distributions of spectrum estimates may be found, and, in particular, confidence intervals for the theoretical spectrum may be derived.

We give a brief nonrigorous description of a small part of stationary time series theory. A more extensive and rigorous mathematical discussion may be found in Anderson (1971), Brillinger (1975), Hannan (1970), or Koopmans (1974).

8.1 STATIONARY TIME SERIES

A (weakly) stationary time series is a collection of random variables $\{X_t\}$ defined for all real t or all integers t, as the case may be, with the following properties:

(i) $E(X_t)$ is constant, and
(ii) $E(X_t X_u)$ depends only on $t - u$.

181

Since $E(X_t)=\mu$ is constant, it may be estimated and subtracted from the series with little effect on what follows. Henceforth we shall assume that $E(X_t)=0$, and state where necessary the modifications to be made if the series mean is subtracted from the data. Property (ii) implies that $\text{var} X_t$ is constant and that

$$\gamma_r = \text{cov}(X_t, X_{t+r}) = \gamma_{-r}$$

does not depend on t. The quantity γ_r is the *theoretical autocovariance at lag r* of $\{X_t\}$ and may be estimated by

$$c_r^* = \frac{1}{n} \sum_{t=|r|}^{n-1} (X_t - \mu)(X_{t-|r|} - \mu), \tag{1}$$

which for $\mu=0$ reduces to $(1/n)\sum X_t X_{t-|r|}$. If μ is replaced by its estimate $\overline{X}$ in (1), we obtain the sample autocovariance c_r used in Chapter 7. The expectation of c_r^* is $(1-|r|/n)\gamma_r$, and hence it is a biased estimate of γ_r, although the bias is small for large n (see Exercise 8.1).

The sequence $\{\gamma_r\}$ is *nonnegative definite*, in the sense that for any constants $\{a_1, \ldots, a_k\}$

$$\sum_{r,s} a_r \gamma_{r-s} a_s \geq 0, \tag{2}$$

for the sum is the variance of a linear combination of random variables in the series and hence is certainly nonnegative. A theorem due to Herglotz states that for any such sequence there exists a nondecreasing function $F(\omega)$ such that

$$\gamma_r = \int_{-\pi}^{\pi} \exp(ir\omega)\, dF(\omega).$$

For a proof see Doob (1953, p. 474). If $F(\omega)$ has a derivative $f(\omega)$, then

$$\gamma_r = \int_{-\pi}^{\pi} \exp(ir\omega) f(\omega)\, d\omega. \tag{3}$$

The functions $F(\omega)$ and $f(\omega)$ are the *spectral distribution function* and *spectral density function*, respectively. The spectral density function is the theoretical counterpart of the (empirical) spectrum defined in Chapter 7. Equation 3 shows that $\{\gamma_r\}$ are the Fourier coefficients of $f(\omega)$. Hence, under mild conditions on $f(\omega)$, the latter may be represented as the Fourier series

$$f(\omega) = \frac{1}{2\pi} \sum_{r=-\infty}^{\infty} \gamma_r \exp(-ir\omega)$$

$$= \frac{1}{2\pi} \left(\gamma_0 + 2 \sum_{r=1}^{\infty} \gamma_r \cos r\omega \right).$$

The equation $\mathrm{var}(X_t) = \gamma_0 = F(\pi) - F(-\pi) = \int f(\omega) \, d\omega$ shows that the variance of the series may be regarded as the sum of components associated with each frequency in the interval $(-\pi, \pi)$, and also that $f(\omega)$ has a finite integral.

Since the autocovariances $\{\gamma_r\}$ are real, it follows that the spectral distribution function $F(\omega)$ is symmetric about zero, in the sense that

$$F(\omega_1) - F(\omega_2) = -\{F(-\omega_1) - F(-\omega_2)\},$$

and that the spectral density function $f(\omega)$ is symmetric in the usual sense.

(i)   A Sinusoid

If

$$X_t = A \cos \omega t + B \sin \omega t,$$

where $E(A) = E(B) = E(AB) = 0$, and $E(A^2) = E(B^2) = a^2$, say, then $\{X_t\}$ is weakly stationary with $E(X_t) = 0$ and

$$\gamma_r = a^2 \cos \omega r.$$

Thus $F(\omega)$ is a step function with jumps of $\frac{1}{2}a^2$ at $-\omega$ and ω. If we add several series of the same kind, with various frequencies and amplitudes, the result is stationary if all the coefficients are uncorrelated. The spectral distribution function then has jumps at each of the frequencies present.

Notice that Fourier analysis of such a series would allow us to obtain estimates of the coefficients A and B, and hence of the amplitude $A^2 + B^2$. However, the parameter a^2 is the expectation of this quantity, and in an arbitrarily long record we still have only one observation each on A and B. Thus, even from a large number of data, we cannot obtain a good estimate of $F(\omega)$ unless $A^2 + B^2$ is nonrandom (has a degenerate distribution). This would happen if, for instance, $A = a\sqrt{2} \cos \Phi$ and $B = -a\sqrt{2} \sin \Phi$, where Φ is uniformly distributed on $(-\pi, \pi)$. This leads to

$$X_t = a\sqrt{2} \cos(\omega t + \Phi),$$

a *random phase* model. Since $F(\omega)$ is not differentiable, there is no spectral density function, except in the sense of generalized functions (δ-functions).

(ii)   White Noise

If $E(X_t)=0$, $E(X_t^2)=\sigma^2$, and $E(X_t X_u)=0$ for $t\neq u$, then $\gamma_0=\sigma^2$ and $\gamma_r=0$, $r\neq 0$. Thus $F(\omega)=\sigma^2\omega/2\pi$. In this case $F(\omega)$ has no discontinuities, and hence the series contains no pure sinusoids as in example (i). The spectral density function is $f(\omega)=\sigma^2/2\pi$, a constant.

(iii)   The Output of a Filter

Suppose that $\{X_t\}$ is stationary with mean zero and spectral distribution function $F_X(\omega)$, and that

$$Y_t = \sum_r g_r X_{t-r}.$$

Then

$$E(Y_t Y_{t+r}) = \sum_{r,s} g_r g_s E(X_{t-r} X_{t+r-s})$$

$$= \sum_{r,s} g_r g_s \int_{-\pi}^{\pi} \exp\{i\omega(h+r-s)\}\, dF_X(\omega)$$

$$= \int_{-\pi}^{\pi} |G(\omega)|^2 \exp(ih\omega)\, dF_X(\omega),$$

where $G(\omega)=\sum g_r \exp(ir\omega)$ is the *transfer function* of the filter (Section 6.2). Thus the spectral distribution function of $\{Y_r\}$ is

$$F_Y(\omega) = \int_{-\pi}^{\omega} |G(\lambda)|^2\, dF_X(\lambda).$$

The function $|G(\omega)|^2$ is called the *power transfer function*.
    The relation between the spectral density functions is simpler, namely,

$$f_Y(\omega) = |G(\omega)|^2 f_X(\omega).$$

Exercise 8.1   *Mean of the Sample Autocovariance*

If c_r^* is defined as in (1), verify that $E(c_r^*)=(1-|r|/n)\gamma_r$. Note that $c_r^*/(1-|r|/n)$ would be an unbiased estimate of γ_r. In the light of the damping used in Chapter 7 to produce spectrum estimates, however, this correction for bias would have little effect, and the effect it had would be in the wrong direction.

Exercise 8.2   *Autocovariances are Nonnegative Definite*

Find the variance of $\sum a_r X_{t-r}$, and hence deduce that $\{\gamma_r\}$ is nonnegative definite.

Exercise 8.3   *(Continuation)   Sample Autocovariances*

Show that the sample autocovariance sequence $\{c_r\}$, where c_r is defined to be zero for $|r| \geq n$, is nonnegative definite. [*Hint*: Use the integral inversion formula

$$c_r = \int_{-\pi}^{\pi} I(\lambda) \exp(ir\lambda) \, d\lambda, \qquad r = 0, \pm 1, \ldots,$$

where $I(\lambda)$ is the periodogram, and the fact that $I(\lambda) \propto |J(\lambda)|^2 \geq 0$.]

Exercise 8.4   *A Sinusoid*

Verify that the sinusoidal series $\{X_t\}$ defined in example (i) is weakly stationary.

Exercise 8.5   *Superposition of Stationary Series*

Suppose that $\{X_t\}$ and $\{Y_t\}$ are weakly stationary, and that $E(X_t Y_u) = E(X_t)E(Y_u)$ for all t and u. Show that $\{Z_t\}$ is weakly stationary, where $Z_t = X_t + Y_t$.

Exercise 8.6   *White Noise*

Show that the spectral distribution function $F(\omega) = \sigma^2 \omega / 2\pi$ corresponds to the autocovariance sequence $\gamma_0 = \sigma^2$, $\gamma_r = 0$ otherwise, and hence to a series of uncorrelated errors.

8.2 CONTINUOUS SPECTRA

We have seen that if a time series contains a periodic component its spectral distribution function has a discontinuity at the corresponding frequency, and hence the spectral density function does not exist. When the spectral distribution function has no discontinuities and the spectral density function exists, the series is said to have a *continuous spectrum*. The autocovariances $\{\gamma_r\}$ of such a series are given by

$$\gamma_r = \int_{-\pi}^{\pi} \exp(ir\omega) f(\omega) d\omega.$$

Since $F(\omega)$ is nondecreasing, we know that $f(\omega) \geqslant 0$, and also that $\int f(\omega) d\omega < \infty$. We now show that any function $f(\omega)$ with these two properties may arise as a spectral density function. Suppose that $G_1(\omega) = f(\omega)^{1/2}$. Then we can find filters with transfer functions arbitrarily close to $G_1(\omega)$. For instance, suppose that $f(\omega)$ is continuous. Then so is $Q_1(x)$, where $Q_1(\cos\omega) = G_1(\omega)$. But in that case we can find a polynomial that is arbitrarily close to $Q_1(x)$, and hence a polynomial in $\cos\omega$ that is arbitrarily close to $G_1(\omega)$. Now it was shown in Section 6.2 that any cosine polynomial is the transfer function of a filter. If we take limits along a sequence of such filters, we arrive at a limit filter, which may involve an infinite number of data points and may therefore be somewhat hypothetical in nature.

Now suppose that $\{U_t\}$ is a white-noise process with variance 2π. If the weights of the limiting filter are $\{g_r\}$, let

$$X_t = \sum g_r U_{t-r}. \tag{4}$$

From example (ii) of Section 8.1 the spectral density of $\{U_t\}$ is the constant 1. From example (iii) of that section, therefore, the spectral density function of $\{X_t\}$ is $|G_1(\omega)|^2 = f(\omega)$. A process of form (4), that is, the output of a filter whose input is white noise, is called a *moving average process*. We deduce that *for any nonnegative function $f(\omega)$ with a finite integral there exists a moving average process whose spectral density function is $f(\omega)$.*

There is another result, which states that *every time series with a continuous spectrum is a moving average process* (Doob, 1953, p. 498). Suppose that the series is $\{X_t\}$, and that its spectral density function is $f(\omega)$. It is now convenient, although not necessary, to assume that $f(\omega)$ is continuous and strictly positive. Then $G_2(\omega) = f(\omega)^{-1/2}$ is bounded and continuous, and as before we may find a limiting filter with transfer function $G_2(\omega)$. Suppose that $\{U_t\}$ is the result of applying this filter to $\{X_t\}$. Then the spectral density function of $\{U_t\}$ is

$$f_U(\omega) = |G_2(\omega)|^2 f(\omega)$$

$$= 1, \qquad -\pi \leqslant \omega \leqslant \pi,$$

and hence $\{U_t\}$ is white noise. Now suppose that we apply to $\{U_t\}$ the filter constructed above with transfer function $G_1(\omega) = f(\omega)^{1/2}$. The result is the same as applying to $\{X_t\}$ a combined filter with transfer function $G_1(\omega)G_2(\omega) = 1$, $-\pi \leqslant \omega \leqslant \pi$. In other words the combined filter is the identity filter, which leaves $\{X_t\}$ unchanged. Thus we have constructed a white-noise series from which $\{X_t\}$ may be obtained by applying a suitable filter, thus proving the result.

It must be remembered that the white-noise process has been defined only in terms of its first and second moments, and that lack of correlation does not imply independence. To see how far from independent the consecutive terms in a white-noise process can be, consider the following example. Suppose that Ω is drawn from a symmetric distribution F on $(-\pi, \pi)$, and that Φ is drawn from the uniform distribution on $(-\pi, \pi)$ and is independent of Ω. Let

$$X_t = a\sqrt{2}\ \cos(\Omega t + \Phi).$$

Conditionally on $\Omega = \omega$, we have the random phase model [example (i) of the preceding section], with mean zero and autocovariances

$$E(X_t X_{t+r} | \Omega = \omega) = a^2 \cos \omega r.$$

The unconditional moments are thus

$$E(X_t X_{t+r}) = a^2 \int_{-\pi}^{\pi} \cos \omega r \, dF(\omega)$$

$$= a^2 \int_{-\pi}^{\pi} \exp(ir\omega) \, dF(\omega),$$

and hence the spectral distribution function is $a^2 F(\omega)$. In particular, if we choose $F(\omega)$ to be linear, $\{X_t\}$ is a white-noise process. However, in any *realization* of the series we see a pure sinusoid!

8.3   TIME AVERAGING AND ENSEMBLE AVERAGING

The spectrum of an empirical series was defined as the aspects of a periodogram that show statistical regularity from one segment of a time series to another. The first spectrum estimate described in Chapter 7 was derived by averaging periodograms of a number of segments. Exercise 7.7 shows that most spectrum estimates may similarly be regarded as the averages over time of local measures of spectral power.

However, the probabilistic model for a stationary time series described in Section 8.1 is one in which an *ensemble* of possible *realizations* of the series is envisaged, and the expectation operator E refers to averaging across the ensemble, rather than along time. A series is said to be *ergodic* if the time average of a quantity is (with probability 1) equal to its ensemble average (see Brillinger, 1975, Section 2.11). Statistical inferences about the structure of a series, such as estimation of its spectrum, should therefore be based on the probabilistic properties of *ergodic* series.

The random phase model [example (i) of Section 8.1] and the white-noise example constructed in Section 8.3 shows that it is easy to construct nonergodic series. In the random phase model, the periodogram $I(\omega; t)$ of the stretch of data $\{X_t, \ldots, X_{t+n-1}\}$ is always equal to $(A^2 + B^2)/2a^2$ times its expectation, and hence so is the time average of $I(\omega; t)$. Thus the series can be ergodic only if $A^2 + B^2$ has a degenerate distribution (that is, if $A^2 + B^2 = 2a^2$ with probability 1). Similarly, any series with discontinuities in its spectral distribution function can be ergodic only if all the amplitudes are (with probability 1) constants. The white-noise series obtained in Section 8.2 by randomizing the frequency illustrates the problem more clearly. From any one realization of this series we would necessarily conclude that the series has a purely discontinuous spectrum, whereas it is in fact continuous. However, a white-noise series in which the observations are *independent* and *identically distributed* is ergodic, and so too is any moving average series constructed from it (such a moving average process is called a *linear process*). In the rest of this chapter we examine the statistical properties of the periodogram of a moving average series, and spectrum estimates constructed from it.

8.4   THE PERIODOGRAM OF A TIME SERIES WITH A CONTINUOUS SPECTRUM

We showed in Chapter 5 that, if $\{U_t\}$ is a Gaussian white-noise series, the periodogram ordinates at the Fourier frequencies (apart from 0 and π) are independently exponentially distributed, with mean $\sigma^2/2\pi$. At 0 (and at π if it is a Fourier frequency) the mean is the same, but the distribution is χ^2 with 1 degree of freedom instead of 2. If the mean is subtracted from the data, the periodogram vanishes identically at frequency 0. If the $\{U_t\}$ are independent with a finite variance but a non-Gaussian distribution, the central limit theorem assures us that as n tends to infinity the joint distribution of any finite number of ordinates approaches the same distribution. Central limit theorems for dependent processes (see, for instance, Ibragimov and Linnik, 1971) show that the same result will hold for dependent white-noise processes, provided that the dependence is limited in time. The rather extreme example of Section 8.2 shows that the result cannot be true for all white-noise processes.

Now suppose that $\{X_t\}$ has a continuous spectrum, with spectral density function $f(\omega)$. Suppose further that

$$X_t = \sum g_r U_{t-r},$$

where $\{U_t\}$ is a white-noise process with variance 2π, either Gaussian or independent, or at least close enough to independent for the exponential distribution to arise as the limit of the distribution of its periodogram. Our argument of Section 6.2 shows that the Fourier transform of a stretch $\{X_0, X_1, \ldots, X_{n-1}\}$ satisfies

$$J_X(\omega) \cong G(\omega)J_U(\omega),$$

and hence the periodogram satisfies

$$I_X(\omega) \cong |G(\omega)|^2 I_U(\omega)$$

$$= f(\omega)I_U(\omega). \tag{5}$$

It is easily shown that $E\{I_U(\omega)\} = 1$, and hence that $E\{I_X(\omega)\} \cong f(\omega)$. Thus $I_X(\omega)$ is an approximately unbiased estimate of $f(\omega)$. However, since its distribution is χ^2 with 2 degrees of freedom, it is a poor estimate and does not improve with increasing series length.

The approximation in (5) is due to end effects of the filter and improves with increasing n. Thus for large n the periodogram of $\{X_t\}$ consists of the spectral density function of $\{X_t\}$ multiplied by the periodogram of a white-noise process. This accounts for the spikiness of the periodograms we obtained in earlier sections. For a rigorous study of the asymptotic behavior of periodograms, see Olshen (1967).

8.5   THE APPROXIMATE MEAN AND VARIANCE OF SPECTRUM ESTIMATES

We are now in a position to derive approximations to the statistical behavior of the spectrum estimates introduced in Chapter 7. The simplest are the discrete spectral averages of periodogram ordinates evaluated at the Fourier frequencies,

$$g(\omega) = \sum_u g_u I(\omega - \omega_u). \tag{6}$$

By the arguments of Section 8.4, such an estimate is a sum of approximately independent and exponentially distributed terms, and thus its approximate distribution may be derived analytically. The following cruder approximation is usually satisfactory.

It was shown in Section 8.4 that

$$EI(\omega) \cong f(\omega).$$

If we assume that the averaging in (6) is over a small interval and that $f(\omega)$ is continuous in that interval, then $f(\omega)$ is approximately constant over the same interval. Hence

$$EI(\omega - \omega_u) \cong f(\omega),$$

and

$$Eg(\omega) \cong f(\omega) \sum g_u.$$

Thus $g(\omega)$ is an approximately unbiased estimate of $f(\omega)$, provided that $\sum g_u = 1$. An approximation to the bias is derived in Section 8.6 (see also Parzen, 1957a).

The variance of an exponentially distributed random variable is the square of its mean, and hence

$$\operatorname{var} I(\omega) \cong f(\omega)^2.$$

It was shown in Section 8.4 that periodogram ordinates at Fourier frequencies are approximately independent, and hence, if $\omega \neq 0$ or π,

$$\operatorname{var} g(\omega) \cong \sum_u g_u^2 f(\omega - \omega_u)^2.$$

If we assume as before that $f(\omega)$ is approximately constant over the interval of averaging, then

$$\operatorname{var} g(\omega) \cong f(\omega)^2 \sum g_u^2. \tag{7}$$

For a lag-weights estimate we write the estimate in the integral form

$$g(\omega) = \int_{-\pi}^{\pi} W(\omega - \lambda) I(\lambda) \, d\lambda, \tag{8}$$

where

$$W(\lambda) = \frac{1}{2\pi} \sum_{|r| < m} w_r \exp(-ir\lambda)$$

is the corresponding spectral window. The integral (8) may be approximated by the sum

$$g_d(\omega) = \frac{2\pi}{n} \sum_j W(\omega - \omega_j) I(\omega_j), \tag{9}$$

which is of form (6), and hence

$$\operatorname{var} g(\omega) \cong \operatorname{var} g_d(\omega)$$

$$\cong \left(\frac{2\pi}{n}\right)^2 f(\omega)^2 \sum_j W(\omega - \omega_j)^2$$

$$\cong \frac{2\pi}{n} f(\omega)^2 \int_{-\pi}^{\pi} W(\lambda)^2 d\lambda. \tag{10}$$

Now

$$\int_{-\pi}^{\pi} W(\lambda)^2 d\lambda = \frac{1}{2\pi} \sum_{|r| < m} w_r^2,$$

and thus an alternative form is

$$\operatorname{var} g(\omega) \cong \frac{1}{n} f(\omega)^2 \sum_{|r| < m} w_r^2.$$

The lag weights are usually given by a formula such as $w_r = w(r/m)$, where $w(x) = 0$ for $|x| \geq 1$. We then find that

$$\sum_{|r| < m} w_r^2 = \sum_{|r| < m} w\left(\frac{r}{m}\right)^2$$

$$\cong m \int_{-1}^{1} w(x)^2 dx.$$

The function $w(x)$ is the *lag window* of the estimate. This gives another approximation,

$$\operatorname{var} g(\omega) \cong \frac{m}{n} f(\omega)^2 \int w(x)^2 dx. \tag{11}$$

This shows that for a given lag window the variance of the spectrum estimate is determined by the ratio m/n. This ratio is just the proportion of sample autocovariances for which the lag weights do not vanish.

Lastly we consider the discrete spectral averages computed from the periodogram ordinates at a set of frequencies $\omega_j' = 2\pi j / n'$ more finely spaced than the Fourier frequencies,

$$g(\omega) = \sum_u g_u I(\omega - \omega_u').$$

If the spectral weights $\{g_u\}$ approximate a smooth window $W(\lambda)$, in the sense that $g_u = (2\pi/n') W(\omega_u')$, then

$$g(\omega) \cong \int W(\lambda) I(\omega - \lambda) d\lambda.$$

(This approximation may be justified by an argument similar to that used in Exercise 8.8.) As for a lag-weights estimate it follows that

$$\operatorname{var} g(\omega) \cong \frac{2\pi}{n} f(\omega)^2 \int_{-\pi}^{\pi} W(\lambda)^2 d\lambda.$$

Now

$$\sum_u g_u^2 = \left(\frac{2\pi}{n'}\right)^2 \sum_u W(\omega_u')^2$$

$$\cong \frac{2\pi}{n'} \int_{-\pi}^{\pi} W(\lambda)^2 d\lambda,$$

and thus

$$\operatorname{var} g(\omega) \cong \frac{n'}{n} f(\omega)^2 \sum g_u^2.$$

The factor n'/n may be regarded as a correction for the finer spacing of the frequencies.

It is often the case that the smooth window $W(\lambda)$ is given by

$$W(\lambda) = mV(m\lambda),$$

where $V(x) = 0$ for $|x| \geqslant \pi$, and

$$\int_{-\pi}^{\pi} V(x) dx = 1.$$

We shall refer to $V(x)$ as a standardized spectral window. By varying m

we obtain a family of windows with different widths. Now

$$\int_{-\pi}^{\pi} W(\lambda)^2 d\lambda = m^2 \int_{-\pi}^{\pi} V(m\lambda)^2 d\lambda$$

$$= m \int_{-\pi}^{\pi} V(x)^2 dx,$$

and hence

$$\operatorname{var} g(\omega) \cong \frac{m}{n} f(\omega)^2 2\pi \int_{-\pi}^{\pi} V(x)^2 dx.$$

Comparison with (11) shows that the parameter m has the same effect as the truncation point of a lag-weights estimate. Note that the span of the resulting filter $\{g_u\}$ is roughly n'/m, and hence its bandwidth (see Section 7.6) is at most $2\pi/m$.

The effect of tapering a set of data before computing the periodogram from which a spectrum estimate is constructed also needs to be taken into account. If the data window is $\{u_t: 0 \leqslant t \leqslant n\}$, then

$$E\{I(\omega)\} \cong U_2 f(\omega),$$

where

$$U_2 = \frac{1}{n} \sum u_t^2.$$

Thus, if we wish to obtain approximately unbiased estimates of spectra, we should normalize the data window so that $U_2 = 1$. (Note that this differs from the normalization that is appropriate for estimating amplitudes of sinusoids. See Exercise 5.6.) With this normalization the effect on the variances of any of the spectrum estimates is to multiply them by a factor that is approximately

$$U_4 = \frac{1}{n} \sum u_t^4$$

(see, for instance, Hannan, 1970, Section V. 4).

In terms of the logarithm of a spectrum estimate, the possible biasing factor U_2 becomes an added constant and is unimportant in most situations. The variance of the logarithm is multiplied by U_4/U_2^2, and this of course does not depend on whether or not the added bias is removed.

For the split cosine bell taper described in Section 5.2, we have

$$u_t = u\left(\frac{2t+1}{2n}\right),$$

where

$$u(x) = \begin{cases} \frac{1}{2}(1-\cos 2\pi x/p) & 0 \leqslant x \leqslant p/2, \\ 1, & p/2 \leqslant x \leqslant 1-p/2, \\ \frac{1}{2}[1-\cos\{2\pi(1-x)/p\}], & 1-p/2 \leqslant x \leqslant 1, \end{cases}$$

and p is the total proportion of the series tapered ($p = 2m/n$, where m is as in equation 9 of Chapter 5). It follows that

$$U_2 \cong 1-p+p\int_0^1 \left\{\tfrac{1}{2}(1-\cos 2\pi x)\right\}^2 ds$$

$$= 1-p+\frac{3p}{8}$$

$$= 1-\frac{5p}{8},$$

$$U_4 \cong 1-p+p\int_0^1 \left\{\tfrac{1}{2}(1-\cos 2\pi x)\right\}^4 dx$$

$$= 1-p+\frac{35p}{128}$$

$$= 1-\frac{93p}{128},$$

and hence

$$\frac{U_4}{U_2^2} \cong \frac{1-93p/128}{(1-5p/8)^2}$$

$$= \tfrac{1}{2}\frac{(128-93p)}{(8-5p)^2}.$$

For $p = 0.1$ and 0.2, the levels of tapering used on the sunspot numbers and the wheat-price index, respectively, U_4/U_2^2 has the values 1.055 and 1.116, respectively. These modest increases in variances are a relatively small price to pay for the protection against possible leakage given by tapering in each case.

Combining these factors, we find that the variance of the discrete spectral average

$$g(\omega) = \sum_u g_u I(\omega - \omega'_n)$$

is

$$\operatorname{var} g(\omega) \cong \frac{U_4}{U_2^2} \frac{n'}{n} f(\omega)^2 \sum g_u^2$$

$$= g^2 f(\omega)^2, \qquad \text{say.} \tag{12}$$

These arguments have to be modified for estimates at frequencies 0 and π, for the periodogram is symmetric about both of these frequencies, and thus a centered average at either frequency is the same as a one-sided average of one-half the length. The result is that the variance of $g(0)$ and $g(\pi)$ is twice that given by (12). (This is true whether or not the series mean is subtracted before the analysis.)

To construct confidence intervals for $f(\omega)$ or $\ln f(\omega)$ we need a approximation to the distribution of $g(\omega)$. The simplest one is found by noting that, since $g(\omega)$ is approximately a sum of independent variables, its distribution is approximately normal. Since its mean is positive and its variance is small, $\ln g(\omega)$ is also approximately normally distributed. A Taylor series expansion of $\ln g(\omega)$ about $f(\omega)$ gives

$$\ln g(\omega) \cong \ln f(\omega) + \frac{g(\omega) - f(\omega)}{f(\omega)},$$

and hence

$$E \ln g(\omega) \cong \ln f(\omega),$$

$$\operatorname{var} \ln g(\omega) \cong \frac{\operatorname{var} g(\omega)}{f(\omega)^2}$$

$$\cong g^2.$$

Thus, for example, a 95% approximate confidence interval for $\ln f(\omega)$ is

$$\ln g(\omega) \pm 1.96g,$$

and for $\log_{10} f(\omega)$ the interval is

$$\log_{10} g(\omega) \pm 1.96 g \log_{10} e.$$

When g^2 is not small, the normal distribution may give a poor approximation to the distribution of $\ln g(\omega)$. Since $g(\omega)$ is approximately a sum of independent exponentially distributed quantities, another possibility is to use the χ^2-distribution as an approximation to that of $g(\omega)$ (see, for instance, Jenkins and Watts, 1968, pp. 87, 252–255). The mean μ, variance σ^2, and degrees of freedom ν of the χ^2-distribution are related by $\nu\sigma^2 = 2\mu^2$, or $\nu = 2\mu^2/\sigma^2$. Thus we approximate the distribution of $g(\omega)$ by the χ^2-distribution with

$$\nu = \frac{2f(\omega)^2}{g^2 f(\omega)^2} = \frac{2}{g^2}$$

degrees of freedom. The quantity $2/g^2$ is called the *equivalent number of degrees of freedom* of the estimate $g(\omega)$. To be exact,

$$\chi^2 = \frac{\nu g(\omega)}{f(\omega)}$$

is approximately χ^2-distribution with ν degrees of freedom. Thus, for instance, a 95% confidence interval for $f(\omega)$ is given by

$$\frac{\nu g(\omega)}{\chi_\nu^2(0.975)} \leqslant f(\omega) \leqslant \frac{\nu g(\omega)}{\chi_\nu^2(0.025)},$$

where $\chi_\nu^2(0.025)$ and $\chi_\nu^2(0.975)$ are the 2.5% and 97.5% points of the χ^2-distribution with ν degrees of freedom. In constructing confidence intervals from the χ^2-distribution it is conventional to place equal masses of probability in the two tails. This does not give the shortest interval for a given level of confifence, and the interval is not symmetric about the estimated value on either a linear or a logarithmic scale. However, it does provide the convenient property that, for instance, the upper limit of a 95% confidence interval is also a 97.5% *upper confidence bound*.

The distribution of $g(\omega)$ is in fact exactly χ^2 for the Daniell estimate (to be exact, for the discrete Daniell estimate with $n' = n$ and no tapering, applied to a Gaussian white-noise series). However, for other estimates it is more skewed than the approximating χ^2-distribution. The distribution of $g(\omega)^\alpha$ is less skewed than that of $g(\omega)$ if $\alpha < 1$, and thus a better approximation can be found by matching the first three approximate moments of

$g(\omega)$ to those of $\beta(\chi^2)^{1/\alpha}$, where χ^2 has the χ^2-distribution with ν degrees of freedom. However, this is not an elementary problem, and we have not used intervals based on its solution.

The estimates of the spectrum of the wheat-price index shown in Figure 7.6 and 7.7 are discrete averages of periodogram ordinates on a fine grid, since the data of length 300 were extended by zeros to length 512. The values of Σg_u^2 are 0.117 and 0.0512, respectively. The correction factor for the spacing of frequencies is $512/300 = 1.707$, and the correction factor for tapering (of 20%, or $p = 0.2$) is 1.116. Thus the combined factor $g^2 = (U_4/U_2^2)(n'/n)\Sigma g_u^2$ has the values 0.223 and 0.0986, respectively.

The approximate 95% confidence limits for $\log_{10} g(\omega)$ based on the normal distribution are therefore ± 0.402 and ± 0.267, respectively. The degrees of freedom in the approximating χ^2-distributions are $2/0.223 = 8.97 \cong 9$ and $2/0.0986 = 20.28 \cong 20$. The corresponding confidence intervals for $\log_{10} g(\omega)$ are -0.325 to 0.523 and -0.236 to 0.319. The limits for the first graph are wide, and clearly none of the local fluctuations in that curve represents statistically significant features. The interval based on the χ^2-distribution is quite asymmetric and is 5% wider than the less precise interval given by normal theory. The limits for the second graph are narrower and more symmetric. The χ^2-interval in this case is 4% wider than the normal interval. In both cases the normal theory limits provide very reasonable approximations, at least as a visual guide to the statistical variability in the spectrum estimates.

For the annual sunspot spectra displayed in Figure 7.8 we have $n = 261$, $n' = 512$, $U_4/U_2^2 = 1.055$, and $\Sigma g_u^2 = 0.0678$. The approximate 95% confidence limits based on the normal distribution are therefore ± 0.319, and the limits based on the appropriate χ^2 = distribution (with 14 degrees of freedom) are -0.271 to 0.396. Again the normal theory interval is accurate enough for graphical presentation. Note that the difference between the estimated spectra of the original series and of the square rooted series is nowhere as large as the width of the confidence interval.

Either of the confidence intervals we have described may be plotted as a band surrounding the estimated spectrum. However it is not a *simultaneous confidence band*, which would have the property that the true spectrum is everywhere covered by the band with a given probability, say 0.95. Instead, we should expect the true spectrum to fall outside the band over about 5% of the interval $(0, \pi)$. The results of Woodroofe and Van Ness (1967) would allow us to construct a simultaneous confidence band for the whole of the estimated spectrum. However, it would be wider in each case by a factor of around $\log m$, where m is $2\pi/$(bandwidth of estimate). For relatively short series such as the wheat-price index and the annual sunspot numbers, these wide simultaneous confidence bands are not particularly useful. It also seems likely that the approximation is good only for very long series (see Hannan, 1970, p. 294).

Exercise 8.7   Approximate Variances

Verify the steps leading to the approximation (7) for the variance of an average of periodogram ordinates at Fourier frequencies.

Exercise 8.8   An Approximation

Show that the difference between (8) and (9) is

$$\frac{1}{\pi} \sum_{r=1}^{m} w_r c_{n-r} \cos r\omega.$$

Note that for $r > 0$

$$c_{n-r} = n^{-1} \sum_{t=0}^{r-1} x_t x_{t+n-r},$$

and hence

$$E\left(c_{n-r}^2\right) = n^{-2} \sum_{t,u=0}^{r-1} E\left(x_t x_{t+n-r} x_u x_{u+n-r}\right)$$

$$\leqslant n^{-2} \sum_{t,u=0}^{r-1} \left\{ E\left(x_t^4\right) E\left(x_{t+n-r}^4\right) E\left(x_u^4\right) E\left(x_{u+n-r}^4\right) \right\}^{1/4}$$

$$= \frac{k^2 r^2}{n^2},$$

where $k^2 = E\left(x_t^4\right)$ assumed to be finite). Thus $|c_{n-r}|$ is of the order kr/n in probability, and hence the difference

$$\frac{1}{\pi} \sum_{r=1}^{m} w_r c_r \cos(n-r)\omega$$

is of the order $k/\pi n \sum r|w_r|$ in probability. If $w_r = w(r/m)$, then

$$\sum_{r=1}^{m} r|w_r| = \sum_{r=1}^{m} r\left|w\left(\frac{r}{m}\right)\right|$$

$$\simeq \frac{m}{2} \int_{-1}^{1} |xw(x)| dx,$$

and so the difference is of the order

$$\frac{km}{2\pi n} \int_{-1}^{1} |xw(x)| \, dx$$

in probability. Now the variance of the discrete average spectrum estimate (9) is of the order m/n, and hence the difference between (8) and (9) is negligible.

Exercise 8.9   *The Effect of Tapering*

Suppose that $\{X_t\}$ is a Gaussian white-noise series, that is, $\{X_t\}$ are independent random variables, each with the standard normal distribution. Suppose that the stretch $\{X_0, \ldots, X_{n-1}\}$ is tapered by $\{u_0, \ldots, u_{n-1}\}$, and that a spectrum estimate $g(\omega)$ with lag weights $\{w_r\}$ is computed from the tapered data. Show that

$$g(\omega) = \frac{1}{2\pi n} \sum_{s,t=0}^{n-1} X_s X_t u_s u_t w_{s-t} \exp\{-i(s-t)\omega\}.$$

(i)   Find the mean and variance of $g(\omega)$. [*Hint*: $E(X_s X_t) = \delta_{s,t} = 1$ if $s = t$, 0 otherwise, and $E(X_s X_t X_u X_v) = \delta_{s,t}\delta_{u,v} + \delta_{s,u}\delta_{t,v} + \delta_{s,v}\delta_{t,u}$.]

(ii)   If $u_s = u\{(2s+1)/2n\}$ for some smooth data window $u(x)$, show that for large n

$$Eg(\omega) \cong \frac{1}{2\pi} \int_0^1 u(x)^2 \, dx,$$

and, if $\omega \neq 0$ or π,

$$\mathrm{var}\, g(\omega) \cong \frac{1}{2\pi n} \sum w_r^2 \int_0^1 u(x)^4 \, dx.$$

8.6   PROPERTIES OF SPECTRAL WINDOWS

It was shown in Section 8.5 that the variance of the discrete spectral average

$$g(\omega) = \frac{2\pi m}{n'} \sum_u V(m\omega_u') I(\omega - \omega_u')$$

is approximately

$$\frac{mU_4}{n} f(\omega)^2 2\pi \int V(x)^2 dx,$$

where U_4 is the correction for tapering. It was also shown that $g(\omega)$ is approximately unbiased. To obtain a more exact description of the bias we may approximate $Eg(\omega) - f(\omega)$ by

$$b(\omega) = \frac{2\pi m}{n'} \sum_u V(m\omega_u') f(\omega - \omega_u') - f(\omega).$$

Now if $f(\omega)$ is smooth enough we have

$$f(\omega - \omega_u') \cong f(\omega) - \omega_u' f'(\omega) + \frac{\omega_u'^2}{2} f''(\omega),$$

and hence

$$b(\omega) \cong \frac{2\pi m}{n'} f(\omega) \sum V(m\omega_u')$$

$$- f(\omega) - \frac{2\pi m}{n'} f'(\omega) \sum \omega_u' V(m\omega_u') + \frac{\pi m}{n'} f''(\omega) \sum \omega_u'^2 V(m\omega_u').$$

Now

$$\frac{2\pi m}{n'} \sum V(m\omega_u') \cong \int V(x) dx = 1,$$

and

$$\frac{2\pi m}{n'} \sum \omega_u' V(m\omega_u') \cong \frac{1}{m} \int x V(x) dx,$$

which vanishes if $V(x)$ is symmetric, the usual case. A similar approximation for the last term gives

$$b(\omega) \cong \frac{1}{2} \left\{ \frac{f''(\omega)}{m^2} \right\} \int x^2 V(x) dx.$$

[It should be noted that this argument does not hold for the Bartlett window, since the sidelobes of the Bartlett window decay as x^{-2}. See, for instance, Jenkins and Watts (1968, p. 247) for an account of the bias of the Bartlett estimate.]

Since the approximate bias depends only on $f''(\omega)$, it clearly represents only the bias caused by local "smudging" of the spectrum, and not leakage. As might be expected, the bias is negative at peaks and positive at troughs. Failure to represent the local features of a spectrum was referred to in Section 8.7 as loss of *resolution*. Since loss of resolution is caused by too high a bandwidth, it seems reasonable to define the bandwidth of the window $W(\lambda) = mV(m\lambda)$ to be

$$\frac{1}{m}\left\{\int x^2 V(x)\,dx\right\}^{1/2}.$$

To the present order of approximation the statistical properties of the spectrum estimate $g(\omega)$ depend only on n, m, $f''(\omega)$, $\int V(x)^2\,dx$, and $\int x^2 V(x)\,dx$. It is clear that we would like both of the last two quantities to be small. If we let $V_c(x) = cV(cx)$, then $V_c(x)$ may also be regarded as a standardized spectral window, since

$$\int V_c(x)\,dx = 1.$$

Now

$$\int V_c(x)^2\,dx = c\int V(x)\,dx,$$

$$\int x^2 V_c(x)\,dx = \frac{1}{c^2}\int x^2 V(x)\,dx,$$

and hence

$$\left\{\int V_c(x)^2\,dx\right\}^2 \int x^2 V_c(x)\,dx = \left\{\int V(x)^2\,dx\right\}^2 \int x^2 V(x)\,dx.$$

Since $V(x)$ and $V_c(x)$ represents the same family of windows, the invariant quantity

$$\left\{\int V(x)^2\,dx\right\}^2 \int x^2 V(x)\,dx$$

is a suitable figure of merit of the standardized window $V(x)$, small values being desirable. For nonnegative windows it may be shown that this figure of merit is always at least $9/125 = 0.072$. (The window that attains this value is not particularly desirable; see Exercise 8.10.) For the rectangular window that corresponds most closely to the modified Daniell procedure used in Section 7.6, the value is $1/12 = 0.083$. For the combined window that results from many repeated applications of the Daniell procedure, the value is approximately $1/4\pi = 0.080$. Neither of these values is unacceptably high compared with the minimum. The results of Exercise 8.11 show that with two applications of Daniell smoothing, with one averaging length being twice the other, we get very close to the minimum. However, three applications give a smoother appearance to the curve (see Section 7.6). Values less than 0.072 may be found if $V(x)$ is not constrained to be nonnegative. (There is no positive lower bound in this case.) However, negative sidelobes in a spectral window are generally undesirable, as they may easily lead to negative values of the spectrum estimate.

The values of this figure of merit for some commonly used windows may be computed from values given by Anderson [1971, Table 9.3.4; $\int x^2 V(x)\,dx = 2k,\ \int V(x)^2\,dx = (1/2\pi)\int k(x)^2\,dx$]. For instance, for the widely used Parzen window the value is 0.088.

The bandwidth of a spectral window is sometimes defined as the width of the rectangular window that gives the same variance. For $W(\lambda) = mV(m\lambda)$ this is therefore $\{m\int V(x)^2\,dx\}^{-1}$. However, this definition has the disadvantage that both bandwidth and approximate variance are determined by the same functional of $V(x)$. Thus it provides no way of discriminating between windows of different shapes.

Further discussion of the properties of spectral windows may be found in Parzen (1957b, 1961), Jenkins (1961), or Rosenblatt (1971). The figure of merit described above was also used by Parzen (1957b). The lower bound and the window that attains it are given by Epanechnikov (1969). The inequality was derived in a different statistical context by Hodges and Lehmann (1956).

Exercise 8.10   The Optimal Window

Let

$$V_0(x) = \begin{cases} \frac{3}{4}(1 - x^2), & |x| \leq 1, \\ 0, & \text{otherwise.} \end{cases}$$

(i)   Verify that

$$\int V_0(x)\,dx = 1,$$

and evaluate

$$\int x^2 V_0(x)\,dx \qquad \text{and} \qquad \int V_0(x)^2\,dx.$$

Show that the figure of merit

$$\left\{ \int V_0(x)^2 \, dx \right\}^2 \int x^2 V_0(x) \, dx$$

has the value 9/125.

(ii)   Let $V(x)$ be any window for which

$$V(x) \geqslant 0, \qquad \int V(x) \, dx = 1,$$

and

$$\int x^2 V(x) \, dx = \int x^2 V_0(x) \, dx.$$

Show that

$$\int V(x)^2 \, dx \geqslant \int V_0(x)^2 \, dx,$$

with equality only if $V(x) = V_0(x)$. [*Hint*: Write $V(x) = V_0(x) + \delta(x)$, and note that $\int \delta(x) \, dx = 0$, $\int x^2 \delta(x) \, dx = 0$, and $\delta(x) \geqslant 0$ for $|x| \geqslant 1$.] Note that because this window does not level off as it approaches 0 it would not give particularly smooth results.

Exercise 8.11   The Daniell Window and Combinations

(i)   Evaluate the figure of merit for the standardized Daniell window

$$V(x) = \begin{cases} \frac{1}{2}, & |x| \leqslant 1, \\ 0, & \text{otherwise.} \end{cases}$$

(ii)   Two applications of Daniell smoothing with the same length of averaging in each step correspond to the triangle window

$$V(x) = \begin{cases} 1 - |x|, & |x| \leqslant 1, \\ 0, & \text{otherwise.} \end{cases}$$

Show that the value of the figure of merit is $2/27 = 0.074$.

(iii)   Two applications of Daniell smoothing with different lengths of averaging correspond to the trapezium window

$$V(x) = \begin{cases} \dfrac{1}{1+p}, & |x| \leqslant p, \\ \dfrac{1-|x|}{1-p^2}, & p \leqslant |x| \leqslant 1, \\ 0, & \text{otherwise,} \end{cases}$$

$0 \leqslant p \leqslant 1$, where the ratio of the two averaging lengths is $(1+p):(1-p)$. Show that the value of the figure of merit is

$$\frac{2(1+2p)^2(1-p^4)}{27(1+p)^4(1-p^2)}$$

and that the values for $p = \frac{1}{2}$ and $\frac{1}{3}$ are 0.0731 and 0.0723, respectively. (Note that these correspond to ratios of smoothing lengths of $3:1$ and $2:1$, respectively.)

Exercise 8.14   The Gaussian Window

Any smooth probability density function may be used as a spectral window. The Gaussian window is

$$V(x) = \frac{1}{\sqrt{2\pi}} \exp(-\tfrac{1}{2}x^2).$$

   (i)   Show that $\int x^2 V(x)\,dx = 1$. [*Hint*: Integrate by parts, and use the fact that $\int V(x)\,dx = 1$.]
   (ii)   Show that $\int V(x)^2\,dx = 1/4\pi = 0.0797$.

   Note that repeated application of filters corresponds to convolution of the windows, and hence by the central limit theorem the combined window is approximately Gaussian.

8.7   ALIASING AND THE SPECTRUM

It often happens that the time series under analysis consists of measurements made at a discrete, equally spaced set of times on some phenomenon that is actually evolving continuously (or at least on a much finer time scale.) Thus there may well be oscillations in the behavior of this phenomenon with frequencies higher than the Nyquist frequency associated with the sampling interval (the time between successive observations; see Section 2.5).

   The simplest situation is as follows. Suppose that $X(t)$ is some random process defined for all values of t. The two most common ways of deriving a discrete set of measurements of such a process are as follows:

   (i)   *sampling*: $Y_t = X(t)$, and
   (ii)   *averaging*: $Z_t = \int_{t-1}^{t} X(u)\,du$.

As we shall see, it is sufficient to consider (i).

The spectrum theory of random processes (stochastic processes) defined for a continuous variable t is essentially similar to that described in Section 8.1. The analog of the autocovariance sequence is the autocovariance function,

$$\gamma(\tau) = \text{cov}\{X(t), X(t+\tau)\},$$

which has the representation

$$\gamma(\tau) = \int_{-\infty}^{\infty} \exp(i\tau\omega) dF(\omega).$$

Note that the limits of integration are now infinite, since there is no Nyquist frequency to limit the frequency of oscillations. The finiteness of $\gamma(0)$ requires that $F(\infty) - F(-\infty) < \infty$. It is easily seen that this implies the continuity of $\gamma(\tau)$.

Now, if we sample $X(t)$ according to (i), we find that

$$\gamma_r = \text{cov}(Y_t, Y_{t+r})$$

$$= \text{cov}\{X(t), X(t+r)\}$$

$$= \gamma(r)$$

$$= \int_{-\infty}^{\infty} \exp(ir\omega) dF(\omega).$$

For simplicity suppose that $F(\omega)$ has a derivative $f(\omega)$. Then

$$\gamma_r = \int_{-\infty}^{\infty} \exp(ir\omega) f(\omega) d\omega$$

$$= \sum_{k=-\infty}^{\infty} \int_{(2k-1)\pi}^{(2k+1)\pi} \exp(ir\omega) f(\omega) d\omega$$

$$= \sum_{k=-\infty}^{\infty} \int_{-\pi}^{\pi} \exp\{ir(\omega - 2k\pi)\} f(\omega - 2k\pi) d\omega$$

$$= \sum_{k=-\infty}^{\infty} \int_{-\pi}^{\pi} \exp(ir\omega) f(\omega - 2k\pi) d\omega,$$

since $\exp(-i2k\pi) = 1$. We may interchange the order of summation and integration to yield

$$\gamma_r = \int_{-\pi}^{\pi} \exp(ir\omega) \sum_{k=-\infty}^{\infty} f(\omega - 2k\pi) d\omega,$$

and it follows that the spectral density function of $\{Y_t\}$ is

$$f_Y(\omega) = \sum_{k=-\infty}^{\infty} f(\omega - 2k\pi), \qquad -\pi < \omega \leqslant \pi. \qquad (13)$$

Function 13 is the *aliased spectral density*. It consists of contributions from all frequencies of the form $\omega - 2k\pi$. Since ω is the *principal alias* (see Section 2.5) of all of these frequencies, this is a very natural result.

The process is illustrated for the function $f(\omega) = 1/(1 + \omega^2/\pi^2)$ in Figure 8.1. The upper graph shows this function "folded in" onto the interval

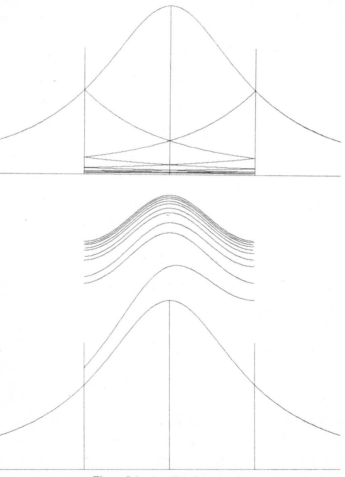

Figure 8.1   An aliased spectrum.

$(-\pi, \pi)$, and the lower graph shows the sum of the different contributions. In this case, the series may be summed to yield $f_Y(\omega) = \pi \sinh \pi / \{2(\cosh \pi - \cos \omega)\}$.

For the series $\{Z_t\}$ obtained by averaging the input series a similar result holds. If we define a new continuous time process

$$W(t) = \int_{t-1}^{t} X(u)\, du, \tag{14}$$

then $\{Z_t\}$ is the result of sampling $\{W(t)\}$. Operation 14 is an example of continuous time filtering and is exactly analogous to the discrete time filtering described in Section 6.2. We again define the transfer function $G(\omega)$ to be the ratio by which a complex sinusoid of frequency ω is changed, and in this case it is

$$\exp\left(\frac{-i\omega}{2}\right)\frac{\sin \omega/2}{\omega/2}.$$

The spectral density function of $\{W_t(t)\}$ is

$$f_W(\omega) = |G(\omega)|^2 f(\omega) = \frac{(\sin \omega/2)^2}{(\omega/2)^2} f(\omega).$$

Since $\{Z_t\}$ is found by sampling $\{W(t)\}$, its spectral density function is given by (13) with $f(\omega)$ replaced by $f_W(\omega)$. Thus

$$f_Z(\omega) = \sum_{k=-\infty}^{\infty} f(\omega - 2k\pi)\frac{\left[\sin\{(\omega - 2k\pi)/2\}\right]^2}{\{(\omega - 2k\pi)/2\}^2}$$

$$= \left(\frac{\sin \omega/2}{\omega/2}\right)^2 \left\{ f(\omega) + \sum_{\substack{k=-\infty \\ k \neq 0}}^{\infty} f(\omega - 2k\pi)\left(\frac{\omega}{\omega - 2k\pi}\right)^2 \right\}. \tag{15}$$

The series in (15) converges more rapidly than that in (13) because of the additional factor, which decays as k^{-2}. Thus aliasing is reduced, but at the cost of some distortion of the leading term. The factor $(\sin \omega/2)^2/(\omega/2)^2$ has the value 1 at $\omega = 0$ but falls to $4/\pi^2 = 0.405$ at $\omega = \pi$. Thus power at frequencies close to π, the Nyquist frequency, is attenuated to around 40% of its true value.

Aliasing can be a confusing phenomenon, since peaks in the true spectrum at frequencies beyond the Nyquist frequency may be strong enough to show in the aliased spectrum. This may give the impression that a frequency is strong when it is not, or the peak may partly obscure another frequency of interest. When the sampling interval (the time interval between consecutive observations) is at our disposal, we may always choose it small enough that no significant power falls outside the Nyquist interval, and thereby avoid aliasing problems completely. However, the cost of collecting and analyzing the data may become prohibitive. One compromise is to collect the data initially with a small sampling interval, filter these data to remove unwanted power at high frequencies, and then *decimate* the result (that is, take every kth observation, where k is chosen to increase the sampling interval to an acceptable value). The simplest such filter would be the discrete analog of (ii), namely, an unweighted average of k successive values. More sophisticated filters could be designed as in Section 6.3 and 6.4. It would be desirable to distort the spectrum as little as possible in the (reduced) Nyquist interval, and to let through a minimum of power from outside that interval. Thus the least squares approximations to a low-pass filter described in Section 6.4 would be appropriate.

9

ANALYSIS OF MULTIPLE SERIES

By a multiple series we mean a number of time series observed simultaneously. This should not be confused with situations in which "time" itself is multidimensional, as when data are collected on a grid in the plane. Fourier analytic methods may also be extended in a fairly obvious way to the latter type of data, but we shall not discuss this subject here. For most of this chapter we examine the situation in which a pair of series is observed, that is, at each time point t we have available a pair of numbers, (x_t, y_t). We shall refer to the separate series $\{x_t\}$ and $\{y_t\}$ as *components*.

When only a single series is observed, the questions that we usually wish to answer are of this form: "What is the *internal* structure of this set of data?" If the data have an oscillatory appearance of the loose form encountered in the examples of preceding chapters, one answer will usually be a description of the data by one of the methods already discussed, namely, harmonic analysis (Chapter 5), complex demodulation (Chapter 6), and spectrum analysis (Chapter 7). When two series are observed, we shall usually be interested in the internal structure of each, but in addition we shall be concerned with their *joint* structure, or the dependence of either series on the other.

Harmonic analysis and its local form, complex demodulation, may be used without modification on a multiple series, by analyzing the series component by component. Suppose, for instance, that we observe a pair of series $\{x_t\}$ and $\{y_t\}$, each containing oscillations at a frequency around λ, but with amplitude and phase fluctuations. Thus we may write

$$x_t = R_{x,t} \cos(\lambda t + \phi_{x,t}) + x_t',$$

$$y_t = R_{y,t} \cos(\lambda t + \phi_{y,t}) + y_t',$$

where $\{x_t'\}$ and $\{y_t'\}$ denote the other components in the two series. By complex demodulation we may find approximations to the instantaneous amplitudes $\{R_{x,t}\}$ and $\{R_{y,t}\}$ and the instantaneous phases $\{\phi_{x,t}\}$ and $\{\phi_{y,t}\}$. From the phases we may also compute the *instantaneous relative phase* of $\{x_t\}$ relative to $\{y_t\}$, $\{\phi_{x,t} - \phi_{y,t}\}$. We may interpret the instantaneous phase of each series as the amount by which its oscillations have drifted from the oscillations in a pure sinusoid with frequency λ, and thus the relative phase measures the extent to which the oscillations in both series drift together. In this way it gives information about the joint behavior of the two series.

Estimated spectra, on the other hand, give us information only about the oscillations in individual series. Similarities between the spectra of two series, such as peaks at similar frequencies, may raise the possibility that the series are related. To investigate such possibilities we may compute estimates of the *cross spectrum* of the two series. This is an extension of the definition of the spectrum and is usually estimated by smoothing the *cross periodogram*.

9.1 THE CROSS PERIODOGRAM

There is a natural extension of the periodogram to multiple series. If we let $J_x(\omega)$ and $J_y(\omega)$ denote the discrete Fourier transforms of the component series, that is,

$$J_x(\omega) = n^{-1} \sum_{t=0}^{n-1} x_t \exp(-it\omega),$$

with $J_y(\omega)$ defined similarly, the periodograms of the components are proportional to $|J_x(\omega)|^2$ and $|J_y(\omega)|^2$, respectively. We shall denote them now by $I_{x,x}(\omega)$ and $I_{y,y}(\omega)$ and shall refer to them as the *autoperiodograms* of the components to distinguish them from the *cross periodogram*,

$$I_{x,y}(\omega) = \frac{n}{2\pi} J_x(\omega) J_y(\omega)^*.$$

Notice that, in contrast with the autoperiodogram, $I_{x,y}(\omega)$ is in general not positive, or even real, and not symmetric. Instead it satisfies the identities

$$I_{x,y}(-\omega) = I_{y,x}(\omega) = I_{x,y}(\omega)^*.$$

If the data were of the form

$$x_t = R_x \exp\{i(\lambda t + \phi_x)\},$$

$$y_t = R_y \exp\{i(\lambda t + \phi_y)\},$$

then we would have

$$J_x(\lambda) = R_x \exp(i\phi_x), \qquad J_y(\lambda) = R_y \exp(i\phi_y)$$

and thus

$$I_{x,y}(\lambda) = \frac{n}{2\pi} R_x R_y \exp\{i(\phi_x - \phi_y)\}.$$

We always have the identity

$$|I_{x,y}(\omega)|^2 = I_{x,x}(\omega)I_{y,y}(\omega),$$

and hence the magnitude of the cross periodogram contains no information that is not in the autoperiodograms. The phase of the cross periodogram in this case is $\phi_x - \phi_y$, the difference of the two phases, or the *relative* phase. Note that the relative phase does not depend on the choice of origin for the time scale. It would be the same for any stretch of n consecutive observations on these series, unlike the phases of the two transforms $J_x(\omega)$ and $J_y(\omega)$ separately.

Thus the crossperiodogram contains no information that is not present in the two Fourier transforms. However, the same is true of the autoperiodogram, and yet we found it a useful tool, especially in a smoothed form. Recall the observation, made in Chapter 7, that when the series being analyzed contains oscillations with no strong periodicity, the periodogram may still show statistical regularity, in that its overall shape, although not its fine structure, is consistent from one stretch of a series to another. For the same kind of data the cross periodogram shows a similar statistical regularity. The smooth function that appears when we average cross periodograms from different stretches of data is called the *cross spectrum*. It is in general complex-valued, since the cross periodogram itself is in general not real-valued.

We shall estimate the cross spectrum by smoothing the cross periodogram, using much the same procedures as for smoothing autoperio-

dograms. The only extra problem that arises with smoothing cross periodograms involves the *alignment* of the series. This is discussed in Section
9.6.

Exercise 9.1   Properties of the Cross Periodogram

(i)   Verify that $I_{x,y}(-\omega) = I_{y,x}(\omega) = I_{x,y}(\omega)^*$.

(ii)   Suppose that $\{x_0, \ldots, x_{n-1}\}$ and $\{y_0, \ldots, y_{n-1}\}$ are shifted cyclically
by amounts h and k, respectively. Show that the phase of the cross
periodogram changes by a linear function of frequency, which vanishes if
$h = k$.

9.2   ESTIMATING THE CROSS SPECTRUM

The formula for the cross periodogram may be rearranged to give

$$I_{x,y}(\omega) = \frac{1}{2\pi} \sum_{|r|<n} c_{x,y,r} \exp(-ir\omega), \tag{1}$$

where

$$c_{x,y,r} = \frac{1}{n} \sum x_t y_{t-r}, \qquad |r| < n,$$

the sum being over all t for which both t and $t-r$ lie in the range
$0, 1, \ldots, n-1$. The quantity $c_{x,y,r}$ is the *cross covariance* of $\{x_t\}$ and $\{y_t\}$ at
lag r. Notice that in general $c_{x,y,r} \neq c_{x,y,-r}$, but $c_{x,y,r} = c_{y,x,-r}$.

Expansion 1 is analogous to the expression given in Section 7.3 for the
autoperiodogram. As a result the discussion of the smoothing of autoperiodograms in Chapter 7 is immediately applicable to the smoothing of
cross periodograms. In particular all the various equivalences derived in
Section 7.5 between the different types of estimates extend to this situation.
We shall use the *discrete spectral average* estimates described in Section 7.5
for smoothing the cross periodogram.

For an example of estimating cross spectra we use the plasma data first
analyzed in Section 6.6. The plasma is generated in a toroidal chamber and
rotates past seven measurement stations equally spaced in a semicircle
around the shorter circumference of the chamber. We examine the first
series, which was used in Section 6.6, the fourth, and the seventh. Figure
6.11 shows the first series, and the others are naturally similar, but
generally displaced by a small amount, corresponding to the time taken for
the plasma to rotate from the first station to the fourth and the seventh,

respectively. For this analysis we use the stretch of data from $t = 1500$ to 1999 (500 observations). This is the stretch of ragged oscillations following the breakdown of the sinusoidal oscillations.

Figure 9.1 shows the base-10 logarithms of estimated spectra of the three series. Each series was corrected for its mean, tapered 20%, and extended to length 512 for analysis. The periodogram of each was then smoothed by three applications of modified Daniell smoothing of length 16 (see Section 7.6; the resulting spectral window is shown inset). An approximate 95% confidence interval for each curve is ± 0.17 (see Section 8.5). The three spectra are generally similar, although only the first, $g_{1,1}(\omega)$, shows a distinct peak. Each shows a fairly sharp decline at around $\omega = 0.3\pi$, followed by a leveling-off or a slight increase. The low power in the third spectrum, $g_{7,7}(\omega)$, at $\omega = \pi$ seems to be inconsistent with the other two spectra. However, the variance of the spectrum estimate at $\omega = \pi$ is twice what it would be at other frequencies, and this might cause the disparity.

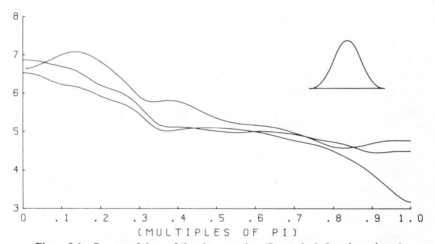

Figure 9.1   Spectra of three of the plasma series. (Spectral window shown inset.)

The cross periodograms $I_{1,4}(\omega)$ and $I_{1,7}(\omega)$ were smoothed in the same way to produce estimates of the cross spectra,

$$g_{1,4}(\omega) = \sum g_u I_{1,4}(\omega - \omega'_u),$$

$$g_{1,7}(\omega) = \sum g_u I_{1,7}(\omega - \omega'_u).$$

These functions have the same symmetries as $I_{x,y}(\omega)$, namely,

$$g_{x,y}(-\omega) = g_{y,x}(\omega) = g_{x,y}(\omega)^*.$$

Like $I_{x,y}(\omega), g_{x,y}(\omega)$ is not in general real.

The real and imaginary parts of the cross spectrum are called the cospectrum and the quadrature spectrum, respectively. The cospectrum measures the extent to which there are oscillations with the same phase in the two series (or with opposite sign, that is, with a phase shift of a half cycle), while the quadrature spectrum measures the extent to which there are oscillations with a phase difference of a quarter cycle (in either direction).

It is ordinarily more useful to consider the magnitude $|g_{x,y}(\omega)|$ and phase $\phi_{x,y}(\omega)$ of the estimated cross spectrum. It is easily shown that if the spectral weights are nonnegative then $|g_{x,y}(\omega)|^2 \leqslant g_{x,x}(\omega)g_{y,y}(\omega)$. Thus it is convenient to consider the ratio

$$s_{x,y}(\omega) = \frac{|g_{x,y}(\omega)|}{\{g_{x,x}(\omega)g_{y,y}(\omega)\}^{1/2}},$$

the estimated *coherency* of $\{x_t\}$ and $\{y_t\}$, which necessarily lies between 0 and 1, rather than the magnitude $|g_{x,y}(\omega)|$ itself. The values 0 and 1 correspond to there being no dependence and complete dependence, respectively, at frequency ω. Notice, however, that this depends on the smoothing. The unsmoothed periodograms satisfy $|I_{x,y}(\omega)|^2 = I_{x,x}(\omega)I_{y,y}(\omega)$, and thus if the coherency is computed without first smoothing the periodograms it is identically 1. [There is some disagreement in the terminology used by different authors. The terms *coherency* and *coherence* have been used variously to describe the quantities $s_{x,y}(\omega)\exp\{i\phi_{x,y}(\omega)\}$, $s_{x,y}(\omega)$, and $s_{x,y}(\omega)^2$. The first is complex, and it is helpful to examine its magnitude and phase separately. Because we are usually more interested in high coherencies, it is convenient to display $s_{x,y}(\omega)^2$, rather than $s_{x,y}(\omega)$ itself. For clarity we shall call this the *squared coherency*.]

The estimated squared coherency and phase for series 1 and 4 are shown in Figure 9.2, and those for series 1 and 7 are shown in Figure 9.3. (The broken lines indicate confidence limits, derived in Section 9.5.) Both squared coherency plots show high coherency at very low frequencies, indicating that there is a low-frequency component common to the three series. They also show high coherency at around $\omega = 0.2\pi$ to 0.3π, a frequency band somewhat higher than that of the peak in $g_{1,1}(\omega)$ (Figure 9.1). Thus the most powerful frequencies are in fact not the frequencies at which the series are most closely related. The high coherency between

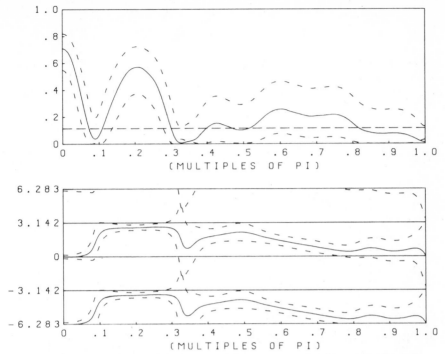

Figure 9.2   Estimated squared coherency and phase spectra for plasma series 1 and 4.

series 1 and 7 at $\omega = \pi$ is presumably related to the low power in $g_{7,7}(\omega)$ at $\omega = \pi$ and may be spurious. If it were real, we would expect it to show also in the squared coherency between series 1 and 4.

To interpret the phase spectrum we must consider the structure of the data. Suppose that the plasma rotates with frequency α. If its cross-sectional structure did not change, the intensity at time t and angle θ would be

$$f(t,\theta) = a(\theta - \alpha t),$$

where $f(0,\theta) = a(\theta)$ is the cross section at time 0. If we expanded $a(\theta)$ in a Fourier series, we would find

$$f(t,\theta) = \sum_r a_r \exp\{ir(\theta - \alpha t)\}.$$

Thus the transforms of the series should contain terms at frequency α and its harmonics. In fact the Fourier coefficients $\{a_r\}$ and the rotational

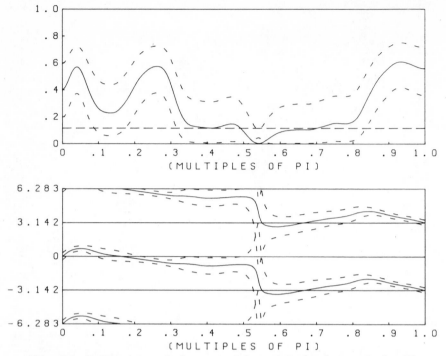

Figure 9.3   Estimated squared coherency and phase spectra for plasma series 1 and 7.

frequency α are time-dependent, and so the autospectra $g_{1,1}(\omega)$, $g_{4,4}(\omega)$, and $g_{7,7}(\omega)$ do not show sharp peaks. The angular separations of station 1 from stations 4 and 7 are a quarter circle and a semicircle, or $\pi/2$ and π radians, respectively. The phase differences at the rth harmonic would thus be $\pi r/2$ and πr, respectively. In the frequency band at which the three series are strongly coherent, from $\omega = 0.2\pi$ to 0.3π, the observed phases are close to π and 0 (or 2π), suggesting that $r = 2$ and that $|\alpha|$ varies between 0.10π and 0.15π. The values $r = 6, 10, \ldots$ would also give these phases, but are less likely. We may verify these conclusions and find the sign of α by performing a similar analysis on the remaining pairs of series.

Exercise 9.2   Rearranging the Cross Periodogram

Verify (1), and write the limits of the sum defining $c_{x,y,r}$. Note that they take different forms according to the sign of r. Verify also that $c_{x,y,r} = c_{y,x,-r}$.

Exercise 9.3   Properties of Estimated Cross Spectra

Show that the estimated cross spectrum $g_{x,y}(\omega)$ has the properties of the cross periodogram mentioned in Exercise 9.1.

Exercise 9.4   The Coherency Inequality

Show that, if the spectral weights are nonnegative, $|g_{x,y}(\omega)|^2 \leqslant g_{x,x}(\omega) \, g_{y,y}(\omega)$.

This is in fact just a special case of the Cauchy-Schwarz inequality. Note that it remains true even if the smoothing procedures used to compute $g_{x,x}(\omega)$, $g_{x,y}(\omega)$, and $g_{y,y}(\omega)$ are not the same, provided that the weights satisfy $g_{x,x,r} \geqslant 0$, $g_{y,y,r} \geqslant 0$, $g_{x,y,r}^2 = g_{x,x,r} g_{y,y,r}$ [here $\{g_{x,x,r}\}$ are the spectral weights used to find $g_{x,x}(\omega)$, etc.].

Exercise 9.5   Invariance of Estimated Coherency under Filtering of Either Component

Suppose that $\{x_t\}$ and $\{y_t\}$ are two series, and that $\{z_t\}$ is obtained from $\{y_t\}$ by a linear filter with transfer function $G(\omega)$. Show that the estimated coherencies $s_{x,y}(\omega)$ and $s_{x,z}(\omega)$ are approximately the same. [*Hint*: assume that $G(\omega)$ is smooth, and in particular is approximately constant over the width of the spectral window. Use the result of Exercise 6.3.]

Exercise 9.6   (Continuation) Coherency as a Measure of Dependence

Show that the estimated coherency of $\{y_t\}$ and $\{z_t\}$ is approximately 1 for all ω. Note that $\{y_t\}$ and $\{z_t\}$ are completely dependent, since each term in the series $\{z_t\}$ may be computed exactly from the terms of $\{y_t\}$.

9.3   THE THEORETICAL CROSS SPECTRUM

We may define the theoretical cross spectrum of a pair of series by making the appropriate extension to the theory described in Chapter 8. The basic extension is to the definition of stationarity.

The stochastic processes $\{X_t\}$ and $\{Y_t\}$ are *jointly weakly stationary* if

  (i) $E(X_t)$ and $E(Y_t)$ are constant, and
 (ii) $E(X_t X_u)$, $E(X_t Y_u)$, and $E(Y_t Y_u)$ depend only on $t - u$.

This implies that both $\{X_t\}$ and $\{Y_t\}$ separately are weakly stationary under the definition of Section 8.1. The extra strength of this definition lies in the requirement that $E(X_t Y_u)$ depend only on $t - u$, since this concerns the joint probabilistic structure of $\{X_t\}$ and $\{Y_t\}$.

For jointly weakly stationary processes $\{X_t\}$ and $\{Y_t\}$, the covariances

$$\gamma_{X,X,r} = \text{cov}(X_t, X_{t-r}),$$
$$\gamma_{X,Y,r} = \text{cov}(X_t, Y_{t-r}),$$

and

$$\gamma_{Y,Y,r} = \mathrm{cov}(Y_t, Y_{t-r})$$

do not depend on t. The first and third are the *autocovariances* of $\{X_t\}$ and $\{Y_t\}$, respectively, while $\gamma_{X,Y,r}$ is the (theoretical) *cross covariance* of $\{X_t\}$ and $\{Y_t\}$ at lag r. Like the empirical cross covariances defined in Section 9.2, the cross covariances are not in general symmetric, but satisfy $\gamma_{X,Y,r} = \gamma_{Y,X,-r}$. These sequences are *jointly nonnegative definite*, in the sense that for any k and constants $a_1,\dots,a_k$ and $b_1,\dots,b_k$,

$$\sum_{r,s} (a_r \gamma_{X,X,r-s} a_s + a_r \gamma_{X,Y,r-s} b_s + b_r \gamma_{Y,X,r-s} a_s + b_r \gamma_{Y,Y,r-s} b_s) \geqslant 0.$$

This implies that in addition to the spectral distribution functions $F_{X,X}(\omega)$ and $F_{Y,Y}(\omega)$, which satisfy

$$\gamma_{X,X,r} = \int_{-\pi}^{\pi} \exp(ir\omega)\, dF_{X,X}(\omega),$$

$$\gamma_{Y,Y,r} = \int_{-\pi}^{\pi} \exp(ir\omega)\, dF_{Y,Y}(\omega),$$

there exists a function $F_{X,Y}(\omega)$ for which

$$\gamma_{X,Y,r} = \int_{-\pi}^{\pi} \exp(ir\omega)\, dF_{X,Y}(\omega).$$

From the properties of the cross covariances, it follows that the *cross spectral distribution function* $F_{X,Y}(\omega)$ satisfies

$$F_{X,Y}(\omega_1) - F_{X,Y}(\omega_2) = \{F_{X,Y}(-\omega_2) - F_{X,Y}(-\omega_1)\}^*$$

$$= \{F_{Y,X}(\omega_1) - F_{Y,X}(\omega_2)\}^*.$$

The spectral distribution functions are nondecreasing (see Section 8.1); this is equivalent to stating that for $\omega_1 > \omega_2$

$$F_{X,X}(\omega_1) - F_{X,X}(\omega_2) \geqslant 0,$$

and

$$F_{Y,Y}(\omega_1) - F_{Y,Y}(\omega_2) \geqslant 0.$$

The function $F_{X,Y}(\omega)$ satisfies the additional condition

$$\{F_{X,X}(\omega_1) - F_{X,X}(\omega_2)\}\{F_{Y,Y}(\omega_1) - F_{Y,Y}(\omega_2)\}$$

$$- |F_{X,Y}(\omega_1) - F_{X,Y}(\omega_2)|^2 \geqslant 0. \tag{2}$$

These three conditions are equivalent to the matrix-valued function

$$\mathbf{F}(\omega) = \begin{bmatrix} F_{X,X}(\omega) & F_{X,Y}(\omega) \\ F_{X,Y}(\omega)^* & F_{Y,Y}(\omega) \end{bmatrix} = \begin{bmatrix} F_{X,X}(\omega) & F_{X,Y}(\omega) \\ F_{Y,X}(\omega) & F_{Y,Y}(\omega) \end{bmatrix}$$

having *nonnegative definite increments*.

If any of these functions is discontinuous at a frequency λ, there is a pure sinusoid of frequency λ in one or both of the component processes. The other important case is that in which $F_{X,X}(\omega)$, $F_{X,Y}(\omega)$, and $F_{Y,Y}(\omega)$ all have derivatives, $f_{X,X}(\omega)$, $f_{X,Y}(\omega)$, and $f_{Y,X}(\omega)$, respectively. Condition 2 is then equivalent to

$$|f_{X,Y}(\omega)|^2 \leqslant f_{X,X}(\omega) f_{Y,Y}(\omega), \tag{3}$$

$-\pi < \omega \leqslant \pi$, or in other words to stating that the matrix-valued function

$$\mathbf{f}(\omega) = \begin{bmatrix} f_{X,X}(\omega) & f_{X,Y}(\omega) \\ f_{X,Y}(\omega)^* & f_{Y,Y}(\omega) \end{bmatrix} = \begin{bmatrix} f_{X,X}(\omega) & f_{X,Y}(\omega) \\ f_{Y,X}(\omega) & f_{Y,Y}(\omega) \end{bmatrix}$$

is *nonnegative definite*. The functions $f_{X,X}(\omega)$ and $f_{Y,Y}(\omega)$ are the spectral density functions of $\{X_t\}$ and $\{Y_t\}$, respectively (see Section 8.1), while $f_{X,Y}(\omega)$ is the *cross-spectral density function*, the theoretical counterpart of the empirical spectrum estimate $g_{X,Y}(\omega)$ described in Section 9.2. It is not in general real-valued, but satisfies

$$f_{X,Y}(-\omega) = f_{Y,X}(\omega) = f_{X,Y}(\omega)^*.$$

The counterparts of the (empirical) coherency and phase are $r_{X,Y}(\omega)$ and $\theta_{X,Y}(\omega)$, where

$$r_{X,Y}(\omega) = \frac{|f_{X,Y}(\omega)|}{\{f_{X,X}(\omega) f_{Y,Y}(\omega)\}^{1/2}},$$

and $\theta_{X,Y}(\omega)$ is defined by

$$f_{X,Y}(\omega) = r_{X,Y}(\omega) \exp\{i\theta_{X,Y}(\omega)\}\{f_{X,X}(\omega) f_{Y,Y}(\omega)\}^{1/2}.$$

From (3) it follows that $r_{X,Y}(\omega)^2 \leqslant 1$. In the next section we derive some statistical properties of the autoperiodogram and cross periodograms in terms of these functions.

Exercise 9.7   Some Operations on Jointly Weakly Stationary Processes

Suppose that $\{X_t\}$ and $\{Y_t\}$ are jointly weakly stationary, with spectral densities $f_{X,X}(\omega)$, $f_{X,Y}(\omega)$, and $f_{Y,Y}(\omega)$.

(i)   Let $Z_t = X_t + Y_t$. Show that $\{Y_t\}$ and $\{Z_t\}$ are jointly weakly stationary, and find the corresponding spectral density functions. [*Hint*: Find the autocovariances and cross covariances, and use the Fourier series for $f_{Y,Z}(\omega)$,

$$f_{Y,Z}(\omega) = \frac{1}{2\pi} \sum \gamma_{y,z,r} \exp(-ir\omega). \Big]$$

(ii)   Let $Z_t = \sum_u g_u X_{t-u}$. Show again that $\{Y_t\}$ and $\{Z_t\}$ are jointly weakly stationary, and find the spectral densities.

(iii)   Suppose that the weights $\{g_u\}$ in (ii) are symmetric about $u = h$. Thus Z_t is the result of applying a symmetric filter and a shift to $\{X_t\}$. Show that the coherency of $\{X_t\}$ and $\{Z_t\}$ is identically 1, and that the phase spectrum is $\theta_{X,Y}(\omega) = h\omega$. [Note that $\{X_t\}$ and $\{Z_t\}$ are jointly weakly stationary by part (ii), since $\{X_t\}$ and $\{X_t\}$ are trivially jointly weakly stationary.]

9.4   THE DISTRIBUTION OF THE CROSS PERIODOGRAM

In this section we derive approximations to the distribution of the auto-periodograms and cross periodograms of a pair of jointly weakly stationary series $\{X_t\}$ and $\{Y_t\}$, with spectral density functions $f_{X,X}(\omega)$, $f_{X,Y}(\omega)$, and $f_{Y,Y}(\omega)$. For rigorous derivations of these results see Anderson (1971, Section 8.2), Brillinger (1975, Section 7.2), or Hannan (1970, Section V.2). We assume first that the series are Gaussian (that is, that all their joint distributions are multivariate normal), and discuss later how this assumption may be relaxed.

The joint distribution of the transforms of two series is simplest if the two series are uncorrelated (that is, if their cross spectrum vanishes). Under the Gaussian assumption this implies independence; hence the two transforms are independent, and the (marginal) distribution of each was found in Section 8.4. We show first that the general distribution may be found from the distribution in this special case. Suppose then that the cross-spectral density $f_{X,Y}(\omega)$ is not identically zero, and let $G(\omega) = f_{X,Y}(\omega)^* / f_{X,X}(\omega)$. Then by an argument similar to that used in Section 8.4

we may find a filter (perhaps only in a limiting sense) with transfer function $G(\omega)$. Suppose that $\{Z_t'\}$ is the result of applying this filter to $\{X_t\}$, and let $Z_t = Y_t - Z_t'$. Then the cross spectrum of $\{X_t\}$ and $\{Z_t\}$ vanishes. The variable Z_t' is in fact the regression of Y_t on $\{X_t: t = 0, \pm 1, \ldots\}$, and Z_t is the residual from this regression. The autospectral density of $\{Z_t\}$ is

$$f_{Z,Z}(\omega) = f_{Y,Y}(\omega) - \frac{|f_{X,Y}(\omega)|^2}{f_{X,X}(\omega)},$$

the *residual spectral density function* of $\{Y_t\}$ after correcting for $\{X_t\}$.

We may now use the argument of Section 8.4 to derive the distribution of $J_X(\omega)$ and $J_Z(\omega)$, the transforms of $\{X_0, \ldots, X_{n-1}\}$ and $\{Z_0, \ldots, Z_{n-1}\}$, respectively. The real and imaginary parts have a joint four-dimensional Gaussian distribution with

$$E\{\mathrm{Re}J_X(\omega)\} = E\{\mathrm{Im}J_X(\omega)\} = E\{\mathrm{Re}J_Z(\omega)\} = E\{\mathrm{Re}J_Z(\omega)\}$$

$$= E\{\mathrm{Re}J_X(\omega)\,\mathrm{Re}J_Z(\omega)\} = E\{\mathrm{Re}J_X(\omega)\,\mathrm{Im}J_Z(\omega)\}$$

$$= E\{\mathrm{Im}J_X(\omega)\,\mathrm{Re}J_Z(\omega)\} = E\{\mathrm{Im}J_X(\omega)\,\mathrm{Im}J_Z(\omega)\} = 0,$$

$$E\{\mathrm{Re}J_X(\omega)\,\mathrm{Im}J_X(\omega)\} \cong E\{\mathrm{Re}J_Z(\omega)\,\mathrm{Im}J_Z(\omega)\} \cong 0,$$

$$\mathrm{var}\{\mathrm{Re}J_X(\omega)\} \cong \mathrm{var}\{\mathrm{Im}J_X(\omega)\} \cong \frac{\pi f_{x,x}(\omega)}{n},$$

$$\mathrm{var}\{\mathrm{Re}J_Z(\omega)\} \cong \mathrm{var}\{\mathrm{Im}J_Z(\omega)\} \cong \frac{\pi f_{z,z}(\omega)}{n}.$$

$$(4)$$

Now $J_Y(\omega)$, the transform of $\{Y_0, \ldots, Y_{n-1}\}$, must satisfy $J_Y(\omega) = J_Z(\omega) + J_{Z'}(\omega)$, and, by the argument of Section 6.2, $J_{Z'}(\omega) \cong G(\omega)J_X(\omega)$, whence

$$J_Y(\omega) \cong J_Z(\omega) + G(\omega)J_X(\omega). \tag{5}$$

The real and imaginary parts of $J_X(\omega)$ and $J_Y(\omega)$ have a 4-dimensional Gaussian distribution because they are linear combinations of the original series, and we may find their moments from (4). They are

$$E\{\mathrm{Re}J_X(\omega)\} = E\{\mathrm{Im}J_X(\omega)\} = E\{\mathrm{Re}J_Y(\omega)\} = E\{\mathrm{Im}J_Y(\omega)\} = 0,$$

$$E\{\mathrm{Re}J_X(\omega)\,\mathrm{Im}J_X(\omega)\} \cong E\{\mathrm{Re}J_Y(\omega)\,\mathrm{Im}J_Y(\omega)\} \cong 0,$$

$$E\{\mathrm{Re}J_X(\omega)\,\mathrm{Re}J_Y(\omega)\} \cong E\{\mathrm{Im}J_X(\omega)J_Y(\omega)\} \cong \frac{\pi c_{X,Y}(\omega)}{h},$$

$$E\left\{\operatorname{Re}J_X(\omega)\operatorname{Im}J_Y(\omega)\right\} \cong -E\left\{\operatorname{Im}J_X(\omega)\operatorname{Re}J_Y(\omega)\right\} \cong \frac{-\pi q_{X,Y}(\omega)}{n},$$

$$\operatorname{var}\left\{\operatorname{Re}J_X(\omega)\right\} \cong \operatorname{var}\left\{\operatorname{Im}J_X(\omega)\right\} \cong \frac{\pi f_{X,X}(\omega)}{n},$$

$$\operatorname{var}\left\{\operatorname{Re}J_Y(\omega)\right\} \cong \operatorname{var}\left\{\operatorname{Im}J_Y(\omega)\right\} \cong \frac{\pi f_{Y,Y}(\omega)}{n}. \tag{6}$$

Here $c_{X,Y}(\omega)$ and $q_{X,Y}(\omega)$ are the cospectrum and quadrature spectrum, respectively [that is, $f_{X,Y}(\omega) = c_{X,Y}(\omega) + iq_{X,Y}(\omega)$].

From the results of Section 8.4 we know that

$$E\left\{I_{X,X}(\omega)\right\} \cong f_{X,X}(\omega), \quad \operatorname{var}\left\{I_{X,X}(\omega)\right\} \cong f_{X,X}(\omega)^2,$$
$$E\left\{I_{Y,Y}(\omega)\right\} \cong f_{Y,Y}(\omega), \quad \operatorname{var}\left\{I_{Y,Y}(\omega)\right\} \cong f_{Y,Y}(\omega)^2.$$

From (6) it follows additionally that

$$E\left\{I_{X,Y}(\omega)\right\} \cong f_{X,Y}(\omega),$$
$$\operatorname{var}\left\{\operatorname{Re}I_{X,Y}(\omega)\right\} \cong \tfrac{1}{2}\left\{f_{X,X}(\omega)f_{Y,Y}(\omega) + c_{X,Y}(\omega)^2 - q_{X,Y}(\omega)^2\right\},$$
$$\operatorname{var}\left\{\operatorname{Im}I_{X,Y}(\omega)\right\} \cong \tfrac{1}{2}\left\{f_{X,X}(\omega)f_{Y,Y}(\omega) - c_{X,Y}(\omega)^2 + q_{X,Y}(\omega)^2\right\},$$
$$\operatorname{cov}\left\{\operatorname{Re}I_{X,Y}(\omega),\operatorname{Im}I_{X,Y}(\omega)\right\} \cong c_{X,Y}(\omega)q_{X,Y}(\omega),$$
$$\operatorname{cov}\left\{\operatorname{Re}I_{X,Y}(\omega),I_{X,X}(\omega)\right\} \cong c_{X,Y}(\omega)f_{X,X}(\omega),$$
$$\operatorname{cov}\left\{\operatorname{Im}I_{X,Y}(\omega),I_{X,X}(\omega)\right\} \cong q_{X,Y}(\omega)f_{X,X}(\omega), \tag{7}$$
$$\operatorname{cov}\left\{\operatorname{Re}I_{X,Y}(\omega),I_{Y,Y}(\omega)\right\} \cong c_{X,Y}(\omega)f_{Y,Y}(\omega),$$
$$\operatorname{cov}\left\{\operatorname{Im}I_{X,Y}(\omega),I_{Y,Y}(\omega)\right\} \cong q_{X,Y}(\omega)f_{Y,Y}(\omega),$$
$$\operatorname{cov}\left\{I_{X,X}(\omega),I_{Y,Y}(\omega)\right\} \cong |f_{X,Y}(\omega)|^2 = c_{X,Y}(\omega)^2 + q_{X,Y}(\omega)^2.$$

Equations 7 give the means, variances, and covariances of the various periodograms at a single frequency ω. As in Section 8.4 we may show that any of these quantities evaluated at a Fourier frequency is approximately independent of any of them evaluated at any other Fourier frequency. This is all the information we need to derive approximate distributions of the smoothed periodograms computed in Section 9.2.

The marginal distributions of $\operatorname{Re}I_{X,Z}(\omega)$ and $\operatorname{Im}I_{X,Z}(\omega)$, the real and imaginary parts of the cross periodogram of the independent series $\{X_t\}$ and $\{Z_t\}$, are both approximately double-exponential (or Laplacian). They are not independent, however, but have an approximately circularly symmetric distribution. The joint distribution of the real and imaginary parts of $I_{X,Y}(\omega)$ is not simple if $f_{X,Y}(\omega) \neq 0$.

These moments and distributions have been found under the assumption of normality. This assumption may be weakened in various ways—for instance, by assuming that all the terms in the white-noise series that generate $\{X_t\}$ and $\{Z_t\}$ are independent of each other (see Section 8.4 for further discussion).

Exercise 9.8 *The Regression of Y_t on $\{X_t\}$*

The regression of Y_t on $\{X_t: t=0, \pm 1,...\}$ is the linear combination of $\{X_t\}$ that is closest to Y_t in the sense of mean squared error.

   (i)   Show that the weights $\{g_u\}$ which minimize

$$E\left(Y_t - \sum g_u X_{t-u}\right)^2$$

satisfy

$$G(\omega) = \sum g_u \exp(-iu\omega) = \frac{f_{X,Y}(\omega)^*}{f_{X,X}(\omega)}.$$

   (ii)   Let $Z_t = Y_t - \sum_u g_u X_{t-u}$, where the weights $\{g_u\}$ are the optimizing weights found in (i). Use the results of Exercise 9.7 to find the spectral densities of $\{X_t\}$ and $\{Z_t\}$. Note that

$$f_{Z,Z}(\omega) = f_{Y,Y}(\omega) - \frac{|f_{X,Y}(\omega)|^2}{f_{X,X}(\omega)}$$

is nonnegative by the coherency inequality (3).

Exercise 9.9 *Moments of Transforms*

Verify that the moments of $J_X(\omega)$ and $J_Y(\omega)$ are given by (6). [*Hint*: Use (4) and (5) and the definition $G(\omega) = f_{X,Y}(\omega)^*/f_{X,X}(\omega) = \{c_{X,Y}(\omega) - iq_{X,Y}(\omega)\}/f_{X,X}(\omega)$].

Exercise 9.10 *Moments of Periodograms*

Verify that the moments of $I_{X,X}(\omega)$, $I_{X,Y}(\omega)$, and $I_{Y,Y}(\omega)$ are given by (7). [*Hint*: If A, B, C, and D have a 4-dimensional Gaussian distribution and $E(A) = E(B) = E(C) = E(D) = 0$, then $E(ABCD) = E(AB)E(CD) + E(AC)E(BD) + E(AD)E(BC)$.]

Exercise 9.11 *Distribution of the Cross Periodogram*

Suppose that A, B, C, and D are independent, and each has the standard normal distribution.

(i)   Show that $AB + CD$ has the double-exponential distribution. [*Hint*: $AB = \frac{1}{4}\{(A+B)^2 - (A-B)^2\}$.]

(ii)   Show that $P = AB + CD$ and $Q = AD - BC$ have a circularly symmetric joint distribution. In other words, if $c^2 + s^2 = 1$, the joint distribution of $cP + sQ$ and $cQ - sP$ is the same as that of P and Q.

9.5   MEANS AND VARIANCES OF SMOOTHED SPECTRA

As with autoperiodograms, we find that cross periodogram ordinates at different Fourier frequencies are uncorrelated. Furthermore, a cross periodogram ordinate at one Fourier frequency is uncorrelated with either autoperiodogram ordinate at a different Fourier frequency, and the same is true for the two autoperiodograms. We may summarize this by saying that any periodogram ordinate (cross or auto-) at a given Fourier frequency is uncorrelated with any periodogram ordinate at any other frequency.

It follows, therefore, that if we calculate a smoothed periodogram using weights $\{g_u\}$, and if the theoretical spectra are approximately constant over the band of frequencies involved, then the resulting spectrum estimates $g_{X,X}(\omega)$, $g_{Y,Y}(\omega)$, and $g_{X,Y}(\omega)$ have the same variances and covariances as $I_{X,X}(\omega)$, $I_{Y,Y}(\omega)$, and $I_{X,Y}(\omega)$ (given at the end of Section 9.4) but multiplied by the factor $\sum g_u^2$. If the data were tapered before the periodogram was calculated, we must multiply this factor by the correction factor U_4/U_2^2 derived in Section 8.5. Lastly, if the periodogram was computed on a finer grid than the Fourier frequencies, we must also multiply by the factor $n'/n > 1$, the ratio of the spacings. In the most general case, the variances and covariances of $g_{X,X}(\omega)$, $g_{X,Y}(\omega)$, and $g_{Y,Y}(\omega)$ are the same as those of $I_{X,X}(\omega)$, $I_{X,Y}(\omega))$, and $I_{Y,Y}(\omega)$ in (7), but multiplied by the factor

$$g^2 = \frac{n'}{n} \frac{U_4}{U_2^2} \sum g_k^2.$$

The phase $\phi_{X,Y}(\omega)$ is given by

$$\phi_{X,Y}(\omega) = \arctan\left\{ \frac{\text{Im}\, g_{X,Y}(\omega)}{\text{Re}\, g_{X,Y}(\omega)} \right\}.$$

In fact we calculate the branch of the arctangent that gives the correct sign to, say, $\text{Re}\, g_{X,Y}(\omega)$, as described in Section 2.2. This may be done using the FORTRAN function ATAN2. We can only obtain approximations to the moments of this by assuming that $g_{X,Y}(\omega)$ has been smoothed enough to ensure that it is close to $f_{X,Y}(\omega)$. If we write $g_{X,Y}(\omega) = f_{X,Y}(\omega) + a + ib$, we may expand $\phi_{X,Y}(\omega)$ in a Taylor series in a and b. The zero- and first-order terms are

$$\theta_{X,Y}(\omega) - \frac{1}{|f_{X,Y}(\omega)|^2}\{aq_{X,Y}(\omega) - bc_{X,Y}(\omega)\},$$

and thus

$$E\{\phi_{X,Y}(\omega)\} \cong \theta_{X,Y}(\omega),$$

$$\text{var}\{\phi_{X,Y}(\omega)\} \cong \frac{g^2}{2}\left\{\frac{1}{r_{X,Y}(\omega)^2} - 1\right\},$$

(8)

where $\theta_{X,Y}(\omega)$ and $r_{X,Y}(\omega)$, defined in Section 9.3, are the theoretical phase and the coherency, respectively. Notice that the approximate variance is large if the coherency is small. The approximation is in fact valid only when it is small compared with π^2, that is, when $r_{X,Y}(\omega)$ is not small.

The estimated phase is approximately normally distributed when this approximation is good, and thus we may construct, say, an approximate 95% confidence interval for $\theta_{X,Y}(\omega)$ as

$$\phi_{X,Y}(\omega) \pm 1.96\, g\left[\frac{1}{2}\left\{\frac{1}{s_{X,Y}(\omega)^2} - 1\right\}\right]^{1/2},$$

(9)

where the theoretical coherency $r_{X,Y}(\omega)$ has been replaced by its estimate $s_{X,Y}(\omega)$. The arguments used by Hannan (1970, Section V.2) suggest that

$$\phi_{X,Y}(\omega) \pm \arcsin\left[t_\nu(0.05)\left[\frac{g^2}{2(1-g^2)}\left\{\frac{1}{s_{X,Y}(\omega)^2} - 1\right\}\right]^{1/2}\right]$$

would be a better approximation, where $t_\nu(\alpha)$ is the $100\alpha\%$ point of the t-distribution with ν degrees of freedom, and $\nu = 2/g^2 - 2$.

We may find the approximate variance of $r_{X,Y}(\omega)$ and the covariance of $r_{X,Y}(\omega)$ and $\phi_{X,Y}(\omega)$ similarly. If $g_{X,X}(\omega) = f_{X,X}(\omega) + c$ and $g_{Y,Y}(\omega) = f_{Y,Y}(\omega) + d$, then, provided that $r_{X,Y}(\omega) \neq 0$,

$$s_{X,Y}(\omega) = \left\{\frac{|g_{X,Y}(\omega)|^2}{g_{X,X}(\omega)g_{Y,Y}(\omega)}\right\}^{1/2}$$

$$\cong r_{X,Y}(\omega) + \frac{ac_{X,Y}(\omega) + bq_{X,Y}(\omega)}{r_{X,Y}(\omega)f_{X,X}(\omega)f_{Y,Y}(\omega)}$$

$$- \frac{1}{2}r_{X,Y}(\omega)\left\{\frac{c}{f_{X,X}(\omega)} + \frac{d}{f_{Y,Y}(\omega)}\right\},$$

(10)

and hence

$$E\{s_{X,Y}(\omega)\} \cong r_{X,Y}(\omega),$$

$$\text{var}\{s_{X,Y}(\omega)\} \cong \frac{g^2}{2}\{1 - r_{X,Y}(\omega)^2\}^2, \tag{11}$$

$$\text{cov}\{s_{X,Y}(\omega), \phi_{X,Y}(\omega)\} \cong 0.$$

The approximate bias in $s_{X,Y}(\omega)$ may be found by taking the expectation of the next term in the Taylor series expansion. The result is

$$\frac{g^2}{4} \frac{\{1 - r_{X,Y}(\omega)^2\}^2}{r_{X,Y}(\omega)},$$

which is always positive. Recall that, if the coherency is computed from the periodograms without smoothing, its value is identically 1. We see that some positive bias remains even after smoothing. There is another source of bias when the spectra are not effectively constant over the bandwidth of the spectral window. Unlike the present source, this increases with the bandwidth. Thus a trade-off is called for in controlling these two sources of bias.

The variance of $s_{X,Y}(\omega)$ depends on $r_{X,Y}(\omega)$ in the same way as that of a correlation coefficient depends on the theoretical correlation. Thus the arctanh transformation makes the variance constant to this order of approximation (see Jenkins and Watts, 1968, p. 379, or Brillinger, 1975, Section 8.9). In fact,

$$E\{\text{arctanh}\, s_{X,Y}(\omega)\} \cong \text{arctanh}\, r_{X,Y}(\omega)$$

and

$$\text{var}\{\text{arctanh}\, s_{X,Y}(\omega)\} \cong \frac{g^2}{2}.$$

Thus we may find, say, a 95% confidence interval for $r_{X,Y}(\omega)$ as $\tanh z_1 \leqslant r_{X,Y}(\omega) \leqslant \tanh z_2$, where z_1 and z_2 are

$$\text{arctanh}\, s_{X,Y}(\omega) \pm \frac{1.96g}{\sqrt{2}} = \tfrac{1}{2}\ln\frac{1 + s_{X,Y}(\omega)}{1 - s_{X,Y}(\omega)} \pm \frac{1.96g}{\sqrt{2}}. \tag{12}$$

Expansion 10 is invalid for $r_{X,Y}(\omega) = 0$, and therefore gives poor approximations for small $r_{X,Y}(\omega)$. The distribution of $s_{X,Y}(\omega)^2$ when $r_{X,Y}(\omega)$

$=0$ is given by

$$\mathrm{pr}\left\{s_{X,Y}(\omega)^2 \leqslant \sigma(p)^2\right\} \cong p,$$

where

$$\sigma(p)^2 = 1 - (1-p)^{g^2/(1-g^2)}$$

(see, for instance, Brillinger, 1975, p. 317). Thus, for example, the 95% point of the distribution is

$$\sigma(0.95)^2 = 1 - 20^{-g^2/(1-g^2)}. \tag{13}$$

Observed values of $s_{X,Y}(\omega)$ less than $\sigma(0.95)$ should therefore be regarded as not significantly different from 0, and confidence interval 12 should be used only if $s_{X,Y}(\omega)$ exceeds this value. Further discussion of the construction of confidence intervals for the estimated phase and coherency may be found in Hannan (1970, Section V.2) and Brillinger (1975, Sections 6.9, 8.9).

The confidence limits described in this section are shown in Figures 9.2 and 9.3. The horizontal broken line in the squared coherency plots is at $\sigma(0.95)^2$, while the broken curves are found from (12). The broken curves on the phase plots are given by (9). The limits become very wide when the squared coherency is small, indicating that the phase is not well determined at these frequencies. In fact the phase is not well determined at any frequency for which $s_{X,Y}(\omega)^2 \leqslant \sigma(0.95)^2$. Thus the rather crude intervals given by (9) are misleadingly narrow for small $s_{X,Y}(\omega)$. The arcsine interval mentioned above would be better behaved in this respect.

It is sometimes of interest to compare the autospectra of a number of series. For instance, we expect the three spectra shown in Figure 9.1 to be the same, except perhaps for a difference in scale. Because the series are not independent, the autospectra are all positively correlated, and this fact has to be taken into account in setting up confidence intervals for differences. (These differences are most naturally calculated as differences of logarithms of spectra or, alternatively, as logarithms of ratios of spectra.) The resulting intervals decrease in width as the coherency increases. Thus we may make very precise comparisons *between* the autospectra of two highly coherent series, even though the autospectrum of each may not be estimated very precisely. The variance of

$$\log_e \frac{g_{X,X}(\omega)}{g_{Y,Y}(\omega)}$$

is approximately $2g\{1-r_{X,Y}(\omega)^2\}$; thus, for example, an approximate 95% confidence interval for

$$\log_e \frac{f_{X,X}(\omega)}{f_{Y,Y}(\omega)}$$

is

$$\log_e \frac{g_{X,X}(\omega)}{g_{Y,Y}(\omega)} \pm 1.96g\sqrt{2\{1-s_{X,Y}(\omega)^2\}} \ ,$$

where $r_{X,Y}(\omega)$ has been replaced by its estimate $s_{X,Y}(\omega)$.

9.6   ALIGNMENT

It was stated in Section 9.2 that smoothing a cross periodogram is essentially no more difficult than smoothing an autoperiodogram, except for the question of *alignment* of the series. In this section we illustrate this problem and discuss ways to avoid it.

Suppose that we observe a pair of series $\{x_t\}$ and $\{y_t\}$, and that in fact $\{y_t\}$ consists of $\{x_t\}$ shifted by h, that is, $y_t = x_{t-h}$. We say that $\{y_t\}$ *lags* $\{x_t\}$ and $\{x_t\}$ *leads* $\{y_t\}$ by h time units. Then the transforms of $\{x_0,\ldots,x_{n-1}\}$ and $\{y_0,\ldots,y_{n-1}\}$ satisfy

$$J_y(\omega) \cong J_x(\omega)\exp(-ih\omega),$$

the approximation being due to h terms at each end of the corresponding sums that are not equal. Thus

$$I_{x,y}(\omega) = \frac{n}{2\pi}J_x(\omega)J_y(\omega)^* \cong I_{x,x}(\omega)\exp(ih\omega).$$

Similarly the cross spectrum of a pair of jointly weakly stationary processes $\{X_t\}$ and $\{Y_t\}$ satisfying the same relationship satisfies

$$f_{x,y}(\omega) = f_{x,x}(\omega)\exp(ih\omega).$$

Then the spectrum estimates should satisfy

$$g_{x,y}(\omega) \cong g_{x,x}(\omega)\exp(ih\omega).$$

However, if h is large enough this cannot be the case.

Consider a lag-weights spectrum estimate with truncation point m,

$$g_{x,y}(\omega) = \frac{1}{2\pi} \sum_{|r|<m} w_r c_{x,y,r} \exp(-ir\omega).$$

Now

$$c_{x,y,r} = n^{-1} \sum x_t y_{t-r}$$

$$\cong n^{-1} \sum x_t x_{t-h-r}$$

$$= c_{x,x,r+h},$$

and hence

$$g_{x,y,r}(\omega) \cong \frac{1}{2\pi} \sum_{|r|<m} w_r c_{x,x,r+h} \exp(-ir\omega). \tag{14}$$

Now $\{c_{x,x,r}: r=0, \pm 1, \ldots, \pm(n-1)\}$ is a symmetric sequence with its maximum at $r=0$. Thus, if $|h| \geqslant m$, the largest term in the sequence $\{c_{x,y,r}\}$ is omitted from the sum (8), and therefore $g_{x,y}(\omega)$ will tend to be smaller than $g_{x,x}(\omega)\exp(ih\omega)$. If h is large enough, all of the large autocovariances are omitted from (8), and thus

$$|g_{x,y}(\omega)| \ll g_{x,x}(\omega) = g_{y,y}(\omega).$$

Hence

$$s_{x,y}(\omega) \cong 0,$$

where

$$s_{x,y}(\omega) = \frac{|g_{x,y}(\omega)|}{\{g_{x,x}(\omega) g_{y,y}(\omega)\}^{1/2}}$$

is the coherency between $\{x_t\}$ and $\{y_t\}$. We would therefore conclude that the two series are unrelated, whereas they are in fact completely dependent.

The phenomenon does not appear only for lag-weights estimates. Any estimate may also be written as a spectral average (see Section 7.5), with an appropriate window. Now $I_{x,x}(\omega)$ is real and nonnegative and hence averages to a positive value. However,

$$I_{x,y}(\omega) \cong I_{x,x}(\omega) \exp(ih\omega),$$

and if the complex factor $\exp(ih\omega)$ goes through one or more complete cycles within the bandwidth of the window, the values of $I_{x,y}(\omega)$ tend to cancel out, and thus their average $g_{x,y}(\omega)$ again may satisfy

$$|g_{x,y}(\omega)| \ll g_{x,x}(\omega).$$

This happens if the bandwidth is at least $2\pi/h$, which is equivalent to $h \geqslant m$ if we define the bandwidth of a lag-weights estimate to be $2\pi/m$.

The simplest way out of this predicament is to realign the data. If we define $\{z_t\}$ by $z_t = y_{t+h}$, there are no alignment problems for $\{x_t, z_t\}$. We have to know h to carry this out, at least to within an error that is small compared with m. The lag of the largest entry in $\{c_{x,y,r}\}$ is the most obvious candidate. A similar correction could be obtained by analyzing $I_{x,y}(\omega) \exp(-ih\omega)$.

In this example, realignment of the two series removes the problem. However, in general it is not possible to find a single realignment that removes the problem at all frequencies. Consider the following example. The series $\{x_t\}$ is the sum of two components $\{u_t\}$ and $\{v_t\}$, where the first has more power at low frequencies, and the second at high frequencies. Now, if $y_t = u_t + v_{t-h}$, then $I_{x,y}(\omega) \cong I_{u,u}(\omega)$ at low frequencies, and $I_{x,y}(\omega) \cong I_{v,v}(\omega) \exp(ih\omega)$ at high frequencies. Thus, whereas no realignment is called for at low frequencies, a realignment of h is needed at high frequencies. More generally, there could be several frequency bands, each requiring its own realignment.

Another way of describing the problem is as follows. Let

$$f_{X,Y}(\omega) = |f_{X,Y}(\omega)| \exp\{i\theta_{X,Y}(\omega)\},$$

and suppose that both $|f_{X,Y}(\omega)|$ and $\theta'_{X,Y}(\omega)$ are smooth, but the latter is large. Then for ω close to λ we have

$$f_{X,Y}(\omega) \cong |f_{X,Y}(\lambda)| \exp[i\{\theta_{X,Y}(\lambda) + (\omega-\lambda)\theta'_{X,Y}(\lambda)\}].$$

Hence

$$E\{g_{X,Y}(\lambda)\} \cong \sum g_k f_{X,Y}\left(\lambda + \frac{2\pi k}{n}\right)$$

$$\cong |f_{X,Y}(\lambda)| \exp\{i\theta_{X,Y}(\lambda)\} \sum_k g_k \exp\left[i\left\{\frac{2\pi k \theta'_{X,Y}(\lambda)}{n}\right\}\right].$$

The first two terms together are $f_{X,Y}(\lambda)$, and the sum is in fact the "lag weight" of "lag" $\theta'_{X,Y}(\lambda)$ (which of course need not be an integer). Thus the problems we have described are associated with rapidly varying phase, in that the sum may be small or even vanish if $\theta'_{X,Y}(\lambda)$ is large enough.

One way to remove the problem therefore would be to modify the periodogram so that its phase cannot vary rapidly. Since

$$f_{X,Y}(\omega) = |f_{X,Y}(\omega)| \exp\{i\theta_{X,Y}(\omega)\},$$

it follows that $I_{X,Y}(\omega)\exp\{-i\theta_{X,Y}(\omega)\}$ is an estimate of a real, positive quantity. Thus there are no more problems in smoothing this function than in smoothing the autoperiodogram. One procedure would be to obtain an initial estimate of the phase spectrum $\theta_{X,Y}(\omega)$, correct $I_{X,Y}(\omega)$ for phase, using this estimate, and then smooth the result to obtain an estimate of $f_{X,Y}(\omega)$. This could then be used to obtain a new estimate of the phase spectrum, thus setting up an iteration. A few steps, say two or three, should be sufficient to achieve reasonable convergence. The problem of alignment is also described by Hannan (1970, Section V.7) and Brillinger (1975, p. 266). Hannan and Thompson (1973) suggest that estimation of the phase spectrum $\theta_{X,Y}(\omega)$ be replaced by estimation of the *group delay* $\theta'_{X,Y}(\omega)$, and discuss the problems involved.

APPENDIX

The following program was used to compute the spectrum estimates shown in this chapter. The transforms of the two series are read in by subprogram DATIN after being computed by a program such as that listed in the Appendix to Chapter 5. The autospectrum estimates, prepared by a program such as that listed in the Appendix to Chapter 7, are similarly read in. Finally the length of the original series, NOBS, is read in so that the normalizing constant may be computed. The input record has the same format as that used by the programs in the Appendix to Chapter 7, to ensure that the same smoothing procedure is used.

NOTE: The program also uses subprograms DATIN (listed in the Appendix to Chapter 2), MODDAN and EXTEND, (in the Appendix to Chapter 7), POLAR (in the Appendix to Chapter 6), and DATOUT (in the Appendix to Chapter 5).

```
C
C     THIS PROGRAM SMOOTHES THE CROSS-PERIODOGRAM
C     OF TWO SERIES.   THE SPECTRUM ESTIMATES ARE COMPUTED
C     BY REPEATED SMOOTHING WITH MODIFIED DANIELL WEIGHTS.
C     THE PROGRAM IS CONTROLLED BY THE
C     FOLLOWING VARIABLES,  WHICH ARE READ IN BY THE PROGRAM.
C
C     NK     THE NUMBER OF SMOOTHING PASSES
C
C     K      AN ARRAY OF THE SMOOTHING PARAMETERS
C
C     NOTE - THE TRANSFORMS AND THE AUTO-SPECTRUM ESTIMATES
C     OF THE TWO SERIES ARE READ IN BY SUBROUTINE DATIN.
C
      DIMENSION SP1(513),SP2(513),TR1(513),TI1(513),
     +     TR2(513),TI2(513),K(10),Y(513)
      DATA PI /3.141593/
      CALL DATIN (TR1,NPGM1,START,STEP,7)
      CALL DATIN (TI1,NPGM1,START,STEP,7)
      CALL DATIN (TR2,NPGM2,START,STEP,8)
      CALL DATIN (TI2,NPGM2,START,STEP,8)
      IF (NPGM1 .EQ. NPGM2) GO TO 20
      WRITE(6,2) NPGM1,NPGM2
2     FORMAT(≠0ERROR - TRANSFORM LENGTHS≠,2I5,≠ DIFFER.≠)
      STOP
20    CONTINUE
      N=NPGM1
      CALL DATIN (SP1,N1,START,STEP,9)
      CALL DATIN (SP2,N2,START,STEP,10)
      IF ((N1 .EQ. N) .AND. (N2 .EQ. N)) GO TO 60
      WRITE(6,3) N,N1,N2
3     FORMAT(≠0ERROR - TRANSFORM LENGTH≠,I5,
     +      ≠ NOT THE SAME AS SPECTRA LENGTHS≠,2I5)
      STOP
60    CONTINUE
      READ(5,1) NOBS,NP2,NK,(K(I),I=1,NK)
1     FORMAT (10X,13I5)
      NP2=(N-1)*2
      CON=FLOAT(NP2**2)/(PI*FLOAT(2*NOBS))
      DO 30 I=1,N
      CR=TR1(I)*TR2(I)+TI1(I)*TI2(I)
      CI=TI1(I)*TR2(I)-TR1(I)*TI2(I)
      TR1(I)=CR*CON
      TI1(I)=CI*CON
30    CONTINUE
```

```
      WRITE(6,6) NK
6     FORMAT(≠0NUMBER OF MODIFIED DANIELL PASSES IS≠,I5/
+          ≠ VALUES OF K ARE -≠)
      WRITE(6,1) (K(I),I=1,NK)
      DO 40 I=1,NK
      CALL MODDAN (TR1,Y,N,K(I), 1.0)
40    CALL MODDAN (TI1,Y,N,K(I),-1.0)
      CALL POLAR (TR1,TI1,N)
      DO 50 I=1,N
50    TR1(I)=TR1(I)/SQRT(SP1(I)*SP2(I))
      START=0.0
      STEP=PI/FLOAT(N-1)
      CALL DATOUT (TR1,N,START,STEP,11)
      CALL DATOUT (TI1,N,START,STEP,11)
      STOP
      END
```

10

FURTHER TOPICS

Many aspects of time series analysis have not been covered in the preceding chapters. Some of these have been omitted because they are not Fourier analysis methods, while others are more advanced and cannot be treated without a more thorough development of the theories of statistics and stochastic processes. This chapter contains brief descriptions of some of these topics, with references to more extensive discussion.

10.1 TIME DOMAIN ANALYSIS

All the methods of analysis described in this book may be termed *frequency domain* methods, in that they seek to describe the fluctuations in one or more series in terms of sinusoidal behavior at various frequencies. The other main type of analysis is *time domain* analysis, in which the behavior of a series is described in terms of the way in which observations at different times are related statistically.

The basic tools in time domain analysis are the sample autocovariances $\{c_r\}$ defined in Section 7.3 and the related sequence of *autocorrelations*, usually defined as

$$r_r = \frac{c_r}{c_0}, \qquad |r| < n.$$

These are a normalized version of the autocovariances and satisfy $r_0 = 1$, $|r_r| \leq 1$. The autocorrelation at lag r, r_r, measures the extent to which the observation at time t is related to the observation at time $t - r$.

234

The autocorrelation sequence and the periodogram $I(\omega)$ [strictly, the normalized periodogram $I(\omega)/c_0$] are equivalent in the sense that each is the Fourier transform of the other. However, the periodogram, either unsmoothed or in its smoothed form as a spectrum estimate, is the simpler of the two to interpret. This is so because periodogram ordinates of a stretch of n observations at frequencies separated by more than $2\pi/n$ are approximately independent, whereas estimated autocorrelations in general have a complicated covariance structure.

Time domain methods have been applied with great success to some specific problems in time series analysis, including *forecasting* and *control*. In the forecasting problem one observes a series $\{x_t\}$ (or, more generally, a number of series, $\{x_t\}$, $\{y_t\}$, ...) up to time n and wishes to predict or forecast some future value x_{n+r}, $r>0$. In the control problem one observes a series $\{x_t\}$ that depends on another series $\{y_t\}$ whose values may be determined by the observer (more generally, $\{x_t\}$ depends on a number of observed series $\{y_t\}$, $\{z_t\}$,..., some of whose values are under the control of the observer). The problem here is to manipulate the controllable series in such a way that future values of $\{x_t\}$ lie as close as possible to desired values. The theory of these problems is described by Whittle (1963), and their solution by time domain methods is discussed extensively by Box and Jenkins (1970).

10.2 SPATIAL SERIES

In the examples used in earlier chapters the variable t has always represented time in one unit or another. However, the methods described are immediately applicable to any set of observations associated with a single variable with constant increments, such as the thickness of a lamina at equally spaced points along a line. More generally, we might have observations at the points of a rectangular grid in the plane (or a higher-dimensional space), say,

$$x_{t,u}, \qquad t=0,\ldots,m-1, \quad u=0,1,\ldots,n-1.$$

Such a collection of data is called a *spatial series*. The basic tool in the Fourier analysis of such data is the multidimensional discrete Fourier transform

$$\begin{aligned}
J_{j,k} &= J(\omega_j,\lambda_k) \\
&= \frac{1}{mn}\sum_{t=0}^{m-1}\sum_{u=0}^{n-1} x_{t,u}\exp\{-i(\omega_j t+\lambda_k u)\},
\end{aligned} \qquad (1)$$

where $\omega_j = 2\pi j/m$ and $\lambda_k = 2\pi k/n$ are Fourier frequencies associated with m and n, respectively. The inverse transform is

$$x_{t,u} = \sum_{j=0}^{m-1} \sum_{k=0}^{n-1} J_{j,k} \exp\{i(\omega_j t + \lambda_k u)\}, \tag{2}$$

which represents the data as linear combinations of elementary sinusoids. As a function of t and u, the (j,k)th term

$$J_{j,k} \exp\{i(\omega_j t + \lambda_k u)\} \tag{3}$$

depends only on $\omega_j t + \lambda_k u$. If we make an orthogonal change of variables to

$$t' = \frac{\omega_j t + \lambda_k u}{(\omega_j^2 + \lambda_k^2)^{1/2}},$$

$$u' = \frac{\lambda_k t - \omega_j u}{(\omega_j^2 + \lambda_k^2)^{1/2}},$$

(3) becomes

$$J_{j,k} \exp(i\alpha t'),$$

where $\alpha = (\omega_j^2 + \lambda_k^2)^{1/2}$, and does not depend on u'. Thus the (j,k)th term is a sinusoidal *surface*, constant along the lines $t' = $ constant, which are parallel to the t'-axis, and with wavelength $2\pi/\alpha = 2\pi/(\omega_j^2 + \lambda_k^2)^{1/2}$ (measured orthogonally to these lines). The inverse transform (2) therefore represents the data as a sum of sinusoidal surfaces with different orientations and wavelengths.

If we were looking for such purely sinusoidal components in the data, we would compute the periodogram

$$I(\omega,\lambda) = \frac{1}{4\pi^2 mn} |J(\omega,\lambda)|^2$$

from a suitably tapered set of data. More commonly, however, we will be interested in a smoothed version

$$g(\omega,\lambda) = \int \int W(\omega - x, \lambda - y) I(x,y) \, dx \, dy$$

or a discrete analog. The simplest way to carry out this smoothing is to divide the (ω,λ) plane into possibly overlapping sets, average each set of periodogram ordinates, and associate the result with the average of the frequencies. An alternative is to divide the data into possible overlapping rectangles, compute periodograms for each, and average them. In terms of the arithmetic operations required, these two approaches are similar, but the latter requires less storage, since the data are processed in segments. Since both m and n may be large, computation time and storage requirements are important considerations in the choice of method.

If more sophisticated smoothing is required, the discrete averaging method may be improved by introducing nonconstant spectral weights and by increasing the number of points at which averages are computed. The equivalent spectral window for the second method is

$$W(\omega,\lambda) = |w(\omega,\lambda)|^2,$$

where

$$w(\omega,\lambda) = \sum w_{t,u} \exp\{-i(t\omega + u\lambda)\},$$

and $\{w_{t,u}\}$ is the data window used on each rectangular segment. Thus any spectral window may be used implicitly by a suitable choice of data window.

Often the orientation of the grid on which the data are collected is arbitrary. In this case it is desirable that the smoothed periodogram be approximately invariant under rotation of the grid (or, rather, *equivariant*, since the orientation of the sinusoidal surfaces is referred to the orientation of the grid). This may be achieved by using a spectral window with circular symmetry. In the second case the spectral window is circularly symmetric if the data window is likewise circularly symmetric.

The statistical properties of the spectrum estimates obtained in this way may be derived by making the appropriate extensions to the definition of a stationary process. Bartlett (1955, pp. 191–197) discusses processes defined in this way, and Rayner (1971) describes the spectrum analysis of spatial series. Unwin and Hepple (1974) review the general analysis of spatial processes and give an extensive bibliography, including several applications of spectrum analysis.

10.3 MULTIPLE SERIES

In Chapter 9 we described the estimation of the cross spectrum, coherency, and phase of a pair of series. These are covariance- and correlation-like quantities, in that they are symmetric (or Hermitian) functions of the two series, and describe the way in which the two series are *related*. Often, however, one series *depends* (or is thought to) on one or more other series, and this dependence is the real object of the investigation. In this case an analog of regression analysis is needed.

Suppose that the dependent series is $\{X_{1,t}\}$, and that it depends on $\{X_{2,t}\}$, $\{X_{3,t}\}$,...,$\{X_{p,t}\}$ through

$$X_{1,t} = \sum_{r=2}^{p} \sum_{u} g_u^{(r)} X_{r,t-u} + Y_t, \tag{4}$$

where Y_t is uncorrelated with all the terms in the independent series. Then the sum in (4) is the linear combination of the terms in the independent series that is closest to $X_{1,t}$ in the sense of mean squared error and is called the *linear regression* of $X_{1,t}$ on the independent series. If $\{X_{2,t}\}$, $\{X_{3,t}\}$,...,$\{X_{p,t}\}$ are jointly weakly stationary (that is, all pairs satisfy the definition in Section 9.1), and $\{Y_t\}$ is weakly stationary, then $\{X_{1,t}\}, \{X_{2,t}\}, \{X_{3,t}\}, \ldots, \{X_{p,t}\}$ are jointly weakly stationary, and we may define the spectra $\{f_{ij}(\omega), i, j = 1, \ldots, p\}$. If $G_i(\omega)$ is the transfer function of the ith filter $\{g_u^{(i)}\}$, it may be shown that

$$
\begin{bmatrix} G_2^*(\omega) \\ \vdots \\ G_p^*(\omega) \end{bmatrix} = \begin{bmatrix} f_{2,2}(\omega) & \cdots & f_{2,p}(\omega) \\ \vdots & & \vdots \\ f_{p,2}(\omega) & \cdots & f_{p,p}(\omega) \end{bmatrix}^{-1} \begin{bmatrix} f_{2,1}(\omega) \\ \vdots \\ f_{p,1}(\omega) \end{bmatrix}.
$$

The residual spectrum $f_{Y,Y}(\omega)$ is given by

$$f_{Y,Y}(\omega) = f_{1,1}(\omega) - \left[f_{1,2}(\omega) \ldots f_{1,p}(\omega) \right] \begin{bmatrix} G_2^*(\omega) \\ \vdots \\ G_p^*(\omega) \end{bmatrix}$$

$$= f_{1,1}(\omega)\{1 - R(\omega)\},$$

say. The quantity $R(\omega)$ is the *multiple coherency* of $\{X_{1,t}\}$ with $\{X_{2,t}\}, \ldots, \{X_{p,t}\}$, and is the analog of the multiple correlation coefficient. Note that if $p = 2$ we have

$$G_2^*(\omega) = \frac{f_{2,1}(\omega)}{f_{2,2}(\omega)}$$

and

$$R(\omega) = \frac{f_{1,2}(\omega) f_{2,1}(\omega)}{f_{1,1}(\omega) f_{2,2}(\omega)}$$

$$= \frac{|f_{1,2}(\omega)|^2}{f_{1,1}(\omega) f_{2,2}(\omega)},$$

the squared coherency.

If it is known that two of the series, say $\{X_{1,t}\}$ and $\{X_{2,t}\}$, depend on the others, we may wish to investigate whether there is any other dependence between them, or whether they are related only by the fact that both depend on $\{X_{3,t}\}, \ldots, \{X_{p,t}\}$. To do this we compute the *partial spectra*

$$\begin{bmatrix} f_{1,1\cdot 3,\ldots,p}(\omega) & f_{1,2\cdot 3,\ldots,p}(\omega) \\ f_{2,1\cdot 3,\ldots,p}(\omega) & f_{2,2\cdot 3,\ldots,p}(\omega) \end{bmatrix}$$

$$= \begin{bmatrix} f_{1,1}(\omega) & f_{1,2}(\omega) \\ f_{2,1}(\omega) & f_{2,2}(\omega) \end{bmatrix} - \begin{bmatrix} f_{1,3}(\omega) & \cdots & f_{1,p}(\omega) \\ f_{2,3}(\omega) & \cdots & f_{2,p}(\omega) \end{bmatrix}$$

$$\times \begin{bmatrix} f_{3,3}(\omega) & \cdots & f_{3,p}(\omega) \\ \vdots & & \vdots \\ f_{p,3}(\omega) & \cdots & f_{p,p}(\omega) \end{bmatrix}^{-1} \begin{bmatrix} f_{3,1}(\omega) & f_{3,2}(\omega) \\ \vdots & \vdots \\ f_{p,1}(\omega) & f_{p,2}(\omega) \end{bmatrix},$$

and hence we may find the *partial coherency* and *phase*. If the partial coherency is not significantly different from 0, the two series $\{X_{1,t}\}$ and $\{X_{2,t}\}$ have no further dependence.

In practice the theoretical spectra are replaced by estimates. Hannan (1970, Chapter V) describes the statistical properties of the resulting estimates of transfer functions, multiple coherencies, and partial coherencies. Brillinger (1975, Chapter 8) gives extensive discussion of the computation and use of these quantities.

Other methods of multivariate analysis such as principal components and canonical correlations may also be extended so as to apply to time series data (see, for instance, Brillinger, 1975, Chapters 9 and 10).

10.4 HIGHER-ORDER SPECTRA

In Section 5.5, it was shown that the *third-order periodogram* can give us information about nonsinusoidal behavior of a series (in that case, the sunspot series). In Section 6.5 a similar quantity was calculated in a local way using complex demodulation. Brillinger and Rosenblatt (1967a, b) define the kth-order autoperiodogram as

$$I_k(\lambda_1,\ldots,\lambda_k) = \left(\frac{n}{2\pi}\right)^{k-1} \prod_{j=1}^{k} J(\lambda_j),$$ (5)

where $\Sigma\lambda_j \equiv 0 \ (mod\,2\pi)$. We have

$$I_2(\omega, -\omega) = \left(\frac{n}{2\pi}\right) J(\omega)J(-\omega)$$

$$= \left(\frac{n}{2\pi}\right)|J(\omega)|^2$$

$$= I(\omega),$$

and thus (5) contains the familiar periodogram as a special case. The definition may be extended in an obvious way to give kth-order cross periodograms of a multiple series.

The expectation of a kth-order periodogram is (approximately) the Fourier transform of the cumulants of the process being analyzed (Brillinger and Rosenblatt, 1967a). Since all cumulants of orders higher than 2 of a Gaussian process vanish, these higher-order spectra provide information about nonnormality of a process. They also give information about nonlinearity in the structure of a process and may be used to indicate whether some transformation such as those discussed in Section 5.7 will lead to a series with simpler structure (see Brillinger, 1965, and Godfrey, 1965).

It is easily seen that nonsinusoidal oscillations in a series are evidence of nonnormality in its probability structure. For a zero-mean Gaussian process $\{X_t\}$ has the symmetry properties that $\{Y_t\}$ and $\{Z_t\}$ have the same distribution as $\{X_t\}$, where $Y_t = -X_t$ and $Z_t = X_{-t}$. However, in the sunspot series (Figure 5.10) we see oscillations that do not have these symmetry properties. In the square roots of the sunspot series (Figure 5.13) the spatial asymmetry has been largely eliminated, but the time asymmetry remains (see also Figure 5.15). Since no transformation of the data can remove this time asymmetry, it represents a fundamental nonnormality in the distribution of the sunspot numbers.

Like spectrum estimates for series defined for multidimensional "time," higher-order spectrum estimates may be calculated either by frequency domain smoothing of the corresponding higher-order periodogram of the whole series, or by time domain averaging of periodograms of segments of the data. Brillinger and Rosenblatt (1967b) describe an estimate of the first type, while Godfrey (1965) uses the second type. The second approach is preferable from a computational point of view. It requires less storage because the data are processed one segment at a time and the periodograms of successive segments may be accumulated as they are computed. Also, for $k > 2$, fewer arithmetic operations are required to compute the periodograms from the transforms.

10.5 NONQUADRATIC SPECTRUM ESTIMATES

All the spectrum estimates discussed in Chapters 7 and 8 are quadratic forms of the data $\{x_t\}$, that is, they may be written as

$$g(\omega) = \sum_{t,u=0}^{n-1} x_t a_{t,u}(\omega) x_u.$$

Similarly the cross spectrum estimates of Chapter 9 are *bilinear forms* of the two series and may be written as

$$g_{x,y}(\omega) = \sum_{t,u=0}^{n-1} x_t a_{t,u}(\omega) y_u.$$

In recent years there has been some interest in other kinds of spectrum estimates. One approach is to assume that the spectrum belongs to some parametric family $f(\omega; \theta_0, \ldots, \theta_p)$, where $\theta_0, \ldots, \theta_p$ are unknown parameters. Estimates $\hat{\theta}_0, \ldots, \hat{\theta}_p$ are computed from the data, and the spectrum is then estimated by

$$g(\omega) = f(\omega; \hat{\theta}_0, \ldots, \hat{\theta}_p).$$

The family used most widely is that of *autoregressive spectra*

$$f(\omega; \theta_0, \ldots, \theta_p) = \frac{\theta_0}{2\pi|1 - \theta_1 \exp(i\omega) - \cdots - \theta_p \exp(ip\omega)|^2} \tag{6}$$

(see, for instance, Parzen, 1969). This is the spectrum of the autoregressive process $\{X_t\}$, defined by

$$X_t = \theta_1 X_{t-1} + \cdots + \theta_p X_{t-p} + U_t, \tag{7}$$

where $\{U_t\}$ is a zero-mean white-noise process with variance θ_0, and

$$E(U_t X_{t-r}) = 0, \qquad r = 1,\ldots,p. \tag{8}$$

The parameters $\theta_0,\ldots,\theta_p$ are constrained to satisfy

  (i) $\theta_0 > 0$, and
  (ii) $1 - \theta_1 z - \cdots - \theta_p z^p \neq 0$ for $|z| \leqslant 1$,

the second condition being required to ensure the stability of the solution to difference equation 7. Relations 7 and 8 show that the autoregressive process has some of the properties of the conventional multiple linear regression model, and its parameters may be estimated in a similar way (see, for instance, Box and Jenkins, 1970, Sections 7.3.1 and A7.5). The exponential model of Bloomfield (1973) may be used in the same way to produce estimated spectra.

Another approach was used by Burg (see, for instance, Lacoss, 1971). Suppose that we know the first m autocovariances of a weakly stationary process (see Section 8.1), $\gamma_0,\ldots,\gamma_{m-1}$ or, alternatively, that we have good estimates of them. These are consistent with any spectrum $f(\omega)$ for which

$$\int_{-\pi}^{\pi} f(\omega)\exp(ir\omega)\,d\omega = \gamma_r, \qquad r = 0,\ldots,m-1, \tag{9}$$

and Burg's procedure is to use the spectrum satisfying (9) that maximizes the *entropy*

$$\int_{-\pi}^{\pi} \log f(\omega)\,d\omega.$$

In this sense it is the smoothest spectrum satisfying (9). It may be shown (Lacoss, 1971) that the resulting *maximum entropy spectrum* is of autoregressive form; in fact, if the m autocovariances are estimated from a stretch of data (the most common case), the resulting estimate is the same as the autoregressive spectrum estimated from the same stretch of data (with $p = m - 1$).

A different estimate, called the *maximum likelihood estimate*, is suggested by Capon (1969). A relationship between this and the maximum entropy spectrum (and hence also the autoregressive spectrum) is given by Burg (1972). Pisarenko (1972) describes a class of generally nonquadratic estimates that includes as special cases the conventional quadratic spectrum estimate (with the Bartlett window) and the maximum likelihood spectrum.

Various arguments have been made in support of the use of these spectrum estimates. The general procedure of assuming a parametric model for a spectrum and then estimating the parameters is efficient (in a statistical sense) when the true spectrum belongs to the family, and is a flexible method if the family is diverse enough. However, the resulting spectrum estimate is a more complex function of the data than a conventional quadratic estimate, and on these grounds is harder to interpret. It is also computationally more complex, at least in the case of a large number of parameters.

It has also been shown that nonquadratic spectra may have greater resolution than a conventional quadratic spectrum estimate using the same number of autocovariances (though we note that some of the comparisons that have been made use the Bartlett estimate as the conventional estimate, and that the Bartlett estimate is undesirable on at least two counts; see Sections 7.3 and 7.5). Nonquadratic spectra would therefore be useful in situations where only a fixed number of *autocovariances* are known or estimated, and a spectrum estimate of the highest possible resolution is needed. However, in the more common case where a stretch of the original *data* is available for analysis, conventional estimates may be constructed with any desired resolution. In the extreme case where it is desired to estimate the frequency of a sinusoid from a few cycles or a fraction of a cycle, the exact least squares methods of Chapter 2 give the maximum resolution.

10.6   INCOMPLETE DATA, IRREGULARLY SPACED DATA, AND POINT PROCESSES

Often a series of observations from which it is desired to compute a spectrum estimate is incomplete (or contains "bad" values, which should be omitted). There are a number of ways in which this situation may be handled.

If the gaps are relatively few and far apart, and especially if each gap is fairly large, the simplest procedure is to treat each uninterrupted stretch of data as a separate series, compute a periodogram for each, and average them. The segments should each be tapered and then extended by zeros to a common length, to ensure that each periodogram is relatively free of leakage and that all are computed at the same frequency.

Suppose that there are m stretches of data, of lengths $n_1, \ldots, n_m$, and that we compute the weighted average

$$g_1(\omega) = \sum_{j=1}^{m} a_j I_j(\omega), \qquad (10)$$

where $a_j > 0$, $j = 1, \ldots, m$, and $\Sigma a_j = 1$. This is computed at the frequencies

$\omega_j' = 2\pi j / n'$ for some $n' \geqslant \max\{n_1, \ldots, n_m\}$ and then smoothed to form

$$g_2(\omega) = \sum_u g_u g_1(\omega - \omega_u')$$

$$= \sum_{j=1}^m a_j \sum_u g_u I_j(\omega - \omega_u'). \tag{11}$$

Now the periodograms of disjoint stretches of data are approximately uncorrelated, and so

$$\operatorname{var} g_2(\omega) \simeq \sum_{j=1}^m a_j^2 \operatorname{var}\left\{ \sum_u g_u I_j(\omega - \omega_u') \right\}, \tag{12}$$

and

$$\operatorname{var}\left\{ \sum_u g_u I_j(\omega - \omega_u') \right\} \simeq f(\omega)^2 \min\left\{ 1, \frac{n'}{n_j} U_j \sum_u g_u^2 \right\}, \tag{13}$$

where U_j is the variance-inflation factor due to tapering of the jth stretch of data (see Section 8.4). Note that the variance-inflation factors may be different for different stretches, since we may wish to taper the same *number of observations*, rather than the same *proportion*, at the ends of each stretch. Expression (13) has been modified so that it does not exceed $f(\omega)^2$, for if n_j were so small that

$$\frac{n'}{n_j} U_j \sum_u g_u^2 > 1, \tag{14}$$

the filtering operation would have little effect on $I_j(\omega)$. Thus

$$\sum_u g_u I_j(\omega - \omega_u') \simeq I_j(\omega),$$

and hence

$$\operatorname{var} \sum_u g_u I_j(\omega - \omega_u') \simeq \operatorname{var} I_j(\omega)$$

$$\simeq f(\omega)^2.$$

From (12) and (13) it follows that $\operatorname{var} g_2(\omega)$ is minimized by taking

$$a_j = \frac{a}{\min\left\{ 1, (n'/n_j) U_j \sum_u g_u^2 \right\}},$$

where a is chosen so that $\Sigma a_j = 1$, that is,

$$\frac{1}{a} = \sum_{j=1}^{m} \frac{1}{\min\left\{ 1, (n'/n_j) U_j \sum g_u^2 \right\}}.$$

Thus the optimal weights in the averaging of $I_j(\omega)$ depend on the amount of subsequent smoothing. Often (14) does not hold for any j, and then

$$a_j = \frac{n_j / U_j}{\sum_{k=1}^{m} (n_k / U_k)}. \tag{15}$$

Thus, if $g_2(\omega)$ is a sufficiently heavily smoothed form of $g_1(\omega)$, the optimal weights are given by (15) and do not depend on the amount of smoothing. If the weights (15) are used when (14) does hold for some j, the effect is to place less weight on the jth segment. Since only the shortest segments are affected in this way, it seems reasonable that weights (15) should be used in any case. The resolution of these short segments is poor, and such down-weighting therefore reduces the bias in $g_2(\omega)$ because of this poor resolution.

On the other hand, if missing or "bad" observations tend to occur in isolated ones or twos, it is simplest to replace any such observation by a linear combination of its neighbors. It may be shown that the effect of this is to introduce a small bias into the spectrum, proportional to the fraction of data that are missing. The effect on the variance of the estimate is to replace $f(\omega)$ by the biased form, and thus the variance of $\log_{10} f(\omega)$ is unaffected.

A combination of these two approaches may also be used. The data are divided into segments at any long gaps, and shorter gaps within each segment are then filled in by linear combinations of their neighbors.

When a moderate proportion of the data is missing, say more than 10% or 20%, it may not be possible to divide the data into stretches with relatively few data points missing in each stretch. In this case even the combined approach may lead to an unacceptable bias in the estimated spectrum. However, in the special case where missing data are replaced by zeros (a trivial linear combination), the bias may be estimated and removed. Estimates constructed in this way have been considered by several authors. Jones (1962) and Parzen (1963) assume that the data are missed in some periodic way. Scheinok (1965) assumes that each observation is observed with a given probability, independently of the others. Bloomfield (1970) generalizes this to a general random mechanism. Jones (1971) makes no assumption about the mechanism that causes observations to be missed, and gives a formula for the approximate variance that differs from the expressions of Scheinok and Bloomfield. However, any of these procedures may result in negative spectrum estimates, which are generally undesirable. This is less likely to happen if the spectrum is close to white than if it has a large dynamic range. Thus the data should usually be prewhitened (see Section 7.6) to some extent, even though this increases the number of missing observations.

It has been assumed throughout this book that the data are collected at equally spaced times. The missing data problems considered above represent a small departure from this assumption. More generally, however, we may wish to estimate the spectrum of a continuous time series $\{x(t)\}$, observed at times $t_1, \ldots, t_n$, which do not fall on a grid. If these epochs of observation are generated by some random mechanism, they are a realization of a *point process*. The estimation of the spectrum of $\{x(t)\}$ is related to the estimation of the spectrum of the point process $\{t_i\}$, which is defined by Bartlett (1963). These estimation problems are discussed by Bartlett and by Brillinger (1972). It is interesting that the first theoretical study of the statistical properties of the periodogram was presented by Schuster (1897) in the context of a point process, the occurrence of earthquakes in Japan.

REFERENCES

Anderson, T. W. (1971). *The Statistical Analysis of Time Series*. New York: Wiley.

Bartlett, M. S. (1948). Smoothing periodograms from time series with continuous spectra. *Nature* **161**, 686–687.

—— (1950). Periodogram analysis and continuous spectra. *Biometrika* **37**, 1–16.

—— (1955). *An Introduction to Stochastic Processes with Special Reference to Methods and Applications*. Cambridge: Cambridge University Press.

—— (1963). The spectral analysis of point processes. *J. R. Stat. Soc., Ser. B* **25**, 264–280.

Bergland, G. D. (1968). A Fast Fourier Transform algorithm using base 8 iterations. *Math. Comput.* **22**, 275–279.

Beveridge, W. H. (1921). Weather and harvest cycles. *Econ. J.* **31**, 429–452.

—— (1922). Wheat prices and rainfall in Western Europe. *J. R. Stat. Soc.* **85**, 412–459.

Bingham, C., M. D. Godfrey, and J. W. Tukey (1967). Modern techniques of power spectrum estimation. *IEEE Trans. Audio Electroacoust.* **AU-15**, 56–66.

Blackman, R. B. and J. W. Tukey (1959). *The Measurement of Power Spectra, from the Point of View of Communications Engineering*. New York: Dover.

Bloomfield, P. (1970). Spectral analysis with randomly missing observations. *J. R. Stat. Soc., Ser. B* **32**, 369–380.

—— (1973). An exponential model for the spectrum of a scalar time series. *Biometrika* **60**, 217–226.

Box, G. E. P. and G. M. Jenkins (1970). *Time Series Analysis: Forecasting and Control*. San Francisco: Holden-Day.

Bray, R. J. and R. E. Loughhead (1964). *Sunspots*. New York: Wiley.

Brent, R. P. (1972). *Algorithms for Minimization without Derivatives*. Englewood Cliffs, N. J.: Prentice-Hall.

Brigham, E. O. (1974). *The Fast Fourier Transform*. Englewood Cliffs, N. J.: Prentice-Hall.

Brillinger, D. R. (1965). An introduction to polyspectra. *Ann. Math. Stat.* **36**, 1351–1374.

—— (1972). The spectral analysis of stationary interval functions. In L. M. LeCam, J. Neyman, and E. L. Scott, Eds., *Proceedings Sixth Berkeley Symposium*. Berkeley: University of California Press. Pp. 483–513.

—— (1973). An empirical investigation of the Chandler wobble and two proposed excitation processes. *Bull. Int. Stat. Inst.* **45**, Book 3, 413–434.

—— (1975). *Time Series: Data Analysis and Theory*. New York: Holt, Rinehart & Winston.

Brillinger, D. R. and M. Rosenblatt (1967a). Asymptotic theory of estimates of k-th order spectra. In B. Harris, Ed., *Spectral Analysis of Time Series*. New York: Wiley. Pp. 153–188.

—— (1967b). Computation and interpretation of k-th order spectra. In B. Harris, Ed., *Spectral Analysis of Time Series*. New York: Wiley. Pp. 189–232.

Burg, J. P (1972). The relationship between maximum entropy spectra and maximum likelihood spectra. *Geophysics* **37**, 375–376.

Capon, J. (1969). High-resolution frequency-wavenumber spectral analysis. *Proc. IEEE* **57**, 1408–1418.

Cooley, J. W., P. A. W. Lewis, and P. D. Welch (1967). Historical notes on the Fast Fourier Transform. *IEEE Trans. Audio Electroacoust.* **AU-15**, 76–79.

Cooley, J. W. and J. W. Tukey (1965). An algorithm for the machine computation of complex Fourier series. *Math. Comput.* **19**, 297–301.

Dale, J. B. (1914a). The resolution of a compound periodic function into simple periodic functions. *Mon. Not. R. Astron. Sos.* **74**, 628–648.

—— (1914b). Note on the number of components of a compound periodic function. *Mon. Not. R. Astron. Soc.* **74**, 664.

Daniell, P. J. (1946). Discussion on the Symposium on Autocorrelation in Time Series. *J. R. Stat. Soc. (Suppl.)* **8**, 88–90.

Doob, J. L. (1953). *Stochastic Processes*. New York: Wiley.

Durbin, J. and G. S. Watson (1950). Testing for serial correlation in least squares regression. I. *Biometrika* **37**, 409–428.

—— (1951). Testing for serial correlation in least squares regression. II. *Biometrika* **38**, 159–178.

—— (1971). Testing for serial correlation in least squares regression. III. *Biometrika* **58**, 1–19.

Epanechnikov, V. A. (1969). Non-parametric estimation of a multivariate probability density. *Theory Probab. Appl.* **14**, 153–158.

Feller, W. (1968). *An Introduction to the Theory of Probability and Its Applications*, Vol. 1, 3rd ed. New York: Wiley.

Fisher, R. A. (1929). Tests of significance in harmonic analysis. *Proc. R. Soc., Ser. A* **125**, 54–59.

Gentleman, W. M. and G. Sande (1966). Fast Fourier Transforms—for fun and profit. In 1966 *Fall Joint Comput. Conf., AFIPS Conf. Proc.* **29**, 563–578.

Godfrey, M. D. (1965). An exploratory study of the bispectrum of an economic time series. *Appl. Stat.* **14**, 48–69.

Good, I. J. (1958). The interaction algorithm and practical Fourier analysis. *J. R. Stat. Soc., Ser. B* **20**, 361–672.

Good, I. J. (1971). The relationship between two Fast Fourier Transforms. *IEEE Trans. Comput.* **C-20**, 310–317.

Granger, C. W. J. and A. O. Hughes (1971). A new look at some old data: the Beveridge wheat price series. *J. R. Stat. Soc., Ser. A* **134**, 413–428.

Grenander, U. and M. Rosenblatt (1953). Statistical spectral analysis of time series arising from stationary stochastic processes. *Ann. Math. Stat.* **24**, 537–558.

—— (1957). *Statistical Analysis of Stationary Time Series*. New York: Wiley.

Hamming, R. W. (1973). *Numerical Methods for Scientists and Engineers*, 2nd ed. New York: McGraw-Hill.

Hamming, R. W. and J. W. Tukey (1949). Measuring noise color. Bell Telephone Laboratories Memorandum.

Hannan, E. J. (1970). *Multiple Time Series*. New York: Wiley.

Hannan, E. J. and P. J. Thompson (1973). Estimating group delay. *Biometrika* **60**, 241–253.

Hart, B. I. and J. von Neumann (1942). Tabulation of the probabilities for the ratio of the mean square successive difference to the variance. *Ann. Math. Stat.* **13**, 207–214.

Hodges, J. L. and E. L. Lehmann (1956). The efficiency of some nonparametric competitors of the t-test. *Ann. Math. Stat.* **27**, 324–335.

Ibragimov, I. A. and Yu. V. Linnik (1971). *Independent and Stationary Sequences of Random Variables*. Groningen: Wolters-Noordhoff.

Jenkins, G. M. (1961). General considerations in the analysis of spectra. *Technometrics* **3**, 133–166.

Jenkins, G. M. and D. G. Watts (1968). *Spectral Analysis and Its Applications*. San Francisco: Holden-Day.

Jones, R. H. (1962). Spectral analysis with regularly missed observations. *Ann. Math. Stat.* **31**, 568–573.

——— (1971). Spectrum estimation with missing observations. *Ann. Inst. Stat. Math.* **23**, 387–398.

Kendall, D. G. (1948). Oscillatory time series. Review of *Contributions to the Study of Oscillatory Time Series,* by M. G. Kendall. *Nature* **161**, 187.

Kendall, M. G. (1971). Studies in the history of probability and statistics. XXVI. The work of Ernst Abbe. *Biometrika* **58**, 369–373.

Knott, C. G. (1897). On lunar periodicities in earthquake frequency. *Proc. R. Soc.* **60**, 457–466.

Koopmans, L. H. (1974). *The Spectral Analysis of Time Series*. New York: Academic.

Lacoss, R. T. (1971). Data adaptive spectral analysis methods. *Geophysics* **36**, 661–675.

Lagrange (1873). Recherches sur la manière de former des tables des planètes d'apres les seules observations. *Oeuvres de Lagrange*, Vol. VI, pp. 507–627.

Lanczos, C. (1961). *Applied Analysis*. Englewood Cliffs, N. J.: Prentice-Hall.

Newton, H. W. (1958). *The Face of the Sun*. Harmondsworth, Middlesex: Penguin.

Olshen, R. A. (1967). Asymptotic properties of the periodogram of a discrete stationary process. *J. Appl. Probab.* **4**, 508–528.

Otnes, R. K. and L. Enochson (1972). *Digital Time Series Analysis*. New York: Wiley.

Parzen, E. (1957a). On consistent estimates of the spectrum of a stationary time series. *Ann. Math. Stat.* **28**, 329–348.

——— (1957b). On choosing an estimate of the spectral density function of a stationary time series. *Ann. Math. Stat.* **28**, 921–932.

——— (1961). Mathematical considerations in the estimation of spectra. *Technometrics* **3**, 167–190.

——— (1963). On spectral analysis with missing observations and amplitude modulation. *Sankhya, Ser. A* **25**, 383–392.

——— (1969). Multiple time series modelling. In P. R. Krishnaiah, Ed., *Multivariate Analysis*, Vol. I. New York: Academic. Pp. 389–409.

Pisarenko, V. F. (1972). On the estimation of spectra by means of nonlinear functions of the covariance matrix. *Geophys. J. R. Astron. Soc.* **28**, 511–531.

——— (1973). Parameter estimation for 2 harmonics with closing frequencies on noise background. *Theory Probab. Appl.* **18**, 826.

Rayner, J. N. (1971). *An Introduction to Spectral Analysis*. London: Pion.

Rosenblatt, M. (1971). Curve estimates. *Ann. Math. Stat.* **42**, 1815–1842.

Scheinok, P. A. (1965). Spectral analysis with randomly missed observations: the binomial case. *Ann. Math. Stat.* **36**, 971–977.

Schuster, A. (1897). On lunar and solar periodicities of earthquakes. *Proc. R. Soc.* **61**, 455–465.

—— (1898). On the investigation of hidden periodicities with application to a supposed 26 day period of meterological phenomena. *Terr. Magn.* **3**, 13–41.

—— (1900). The periodogram of magnetic declination as obtained from the records of the Greenwich Observatory during the years 1871–1895. *Cambridge Phil. Trans.* **18**, 107–135.

—— (1906). On the periodicities of sunspots. *Phil. Trans. R. Soc., Ser. A* **206**, 69–100.

Shimshoni, M. (1971). On Fisher's test of significance in harmonic analysis. *Geophys. J. R. Astron. Soc.* **23**, 373–377.

Slutsky, E. (1937). The summation of random causes as the source of cyclic processes. *Econometrica* **5**, 105–146.

Spar, J. and J. A. Mayer (1973). Temperature trends in New York City: a postscript. *Weatherwise* **26**, 128–130.

Stewart, B. and W. Dodgson (1879). Preliminary report to the Committee on Solar Physics on a method of detecting the unknown inequalities of a series of observations. *Proc. R. Soc.* **29**, 106–122.

Stokes, G. G. (1879). Note on the paper by Stewart and Dodgson. *Proc. R. Soc.* **29**, 122–123.

Thomson, W. (1876). On an instrument for calculating ($\int \phi(x)\psi(x)dx$), the integral of the product of two given functions. *Proc. R. Soc.* **24**, 266–268.

—— (1878). Harmonic analyzer. *Proc. R. Soc.* **27**, 371–373.

Titchmarsh, E. C. (1939). *Theory of Functions*. London: Oxford University Press.

Tukey, J. W. (1967). An introduction to the calculations of numerical spectrum analysis. In B. Harris, Ed., *Spectral Analysis of Time Series*. New York: Wiley. Pp. 25–46.

Unwin, D. J. and L. W. Hepple (1974). The statistical analysis of spatial series. *Statistician* **23**, 211–227.

von Neumann, J. (1941). Distribution of the ratio of the mean square successive difference to the variance. *Ann. Math. Stat.* **12**, 367–395.

—— (1942). A further remark concerning the distribution of the ratio of the mean square successive difference to the variance. *Ann. Math. Stat.* **13**, 86–88.

von Neumann, J., R. H. Kent, H. R. Bellison, and B. I. Hart (1941). The mean square successive difference. *Ann. Math. Stat.* **12**, 153–162.

Waldmeier, M. (1961). *The Sunspot-Activity in the Years* 1610–1960. Zürich: Schulthess.

Walker, A. M. (1971). On the estimation of a harmonic component in a time series with stationary independent residuals. *Biometrika* **58**, 21–36.

Whittaker, E. T. and G. Robinson (1924). *The Calculus of Observations*. London: Blackie and Son.

Whittle, P. (1952). The simultaneous estimation of a time series harmonic components and covariance structure. *Trab. Estad.* **3**, 43–57.

—— (1963). *Prediction and Regulation by Linear Least-Squares Methods*. London: The English Universities Press.

Wold, H. O. A. (1954). *A Study in the Analysis of Stationary Time Series*, 2nd ed. Stockholm: Almqvist and Wiksell.

Woodroofe, M. B. and J. W. Van Ness (1967). The maximum deviation of sample spectral densities. *Ann. Math. Stat.* **38**, 1558–1569.

INDEX